SMP 11-16

Book G9

CAMBRIDGE
UNIVERSITY PRESS

Published by the Press Syndicate of the University of Cambridge
The Pitt Building, Trumpington Street, Cambridge CB2 1RP
40 West 20th Street, New York, NY 10011–4211, USA
10 Stamford Road, Oakleigh, Melbourne 3166, Australia

First published 1994
Reprinted 1995

Produced by Gecko Limited, Bicester, Oxon.

Printed in Great Britain by Scotprint Ltd, Musselburgh, Scotland

A catalogue record for this book is available from the British Library

ISBN 0 521 45749 1 paperback

Contents

Check the rules

A game for two people

Use the top grid on the worksheet.

Aim: to win 3 squares in a row.

Players take it in turns.

Throw the two dice.
Check each square on the grid.
If your throw fits the rule in the square, you win the square.

So this throw . . .

. . . wins this square.

This throw . . .

. . . wins this square.

Number on red dice $+$ Number on black dice $= 7$	Number on red dice $-$ Number on black dice $= 1$	Number
Number on red dice $=$ Number on black dice	Number on red dice $+$ Number on black dice $= 3$	Number
Number on red dice $=$ Number on black dice $+ 2$	Number on red dice $-$ Number on black dice $= 5$	Number

If you win a square, write your name in it.
Take it in turns to throw the dice.

The first player to capture three squares in a line is the winner.
If no-one gets three in a line, play again.

1 Writing algebra

A Shorthand

(You need to play the game opposite before you start this section.)

Look at the rules on worksheet G9–1.

Number on red dice	Number on red dice	Number
+	**–**	.
Number on black dice	Number on black dice	Number
= 7	**= 1**	

There is a lot of writing in the rules.

For example, the rule in the first box is

number on red dice + number on black dice = 7.

You can make the rules much shorter if you write
    r to stand for *the number on the red dice*
and *b* to stand for *the number on the black dice*.

    number on red dice + number on black dice = 7
becomes $r + b = 7$

> **A1** The rule in the second box is
>   *number on red dice – number on black dice = 1.*
>
> Write this in the new shorthand.
> (Use *r* for *number on red dice*
> and *b* for *number on black dice*.)

> **A2** Write *number on red dice + number on black dice = 5*
> in the new shorthand.

> **A3** Write *number on red dice = number on black dice*
> in the new shorthand.

> **A4** Look at the grid at the bottom of worksheet G9–1.
> Write each of the rules in the new shorthand.
> (The first one is done for you.)

> **A5** Play the game again on the bottom grid.

B Machine hire

This machine tells you
how much it costs
to hire a cement mixer.

Number of days hired → ×5 → +8 → Cost of hiring in pounds

You can write the rule much more simply.

Write *n* to stand for *the number of days hired*
$\quad$ *c* to stand for *the cost of hiring in pounds.*

The rule becomes
$$n \times 5 + 8 = c$$

B1 This machine tells you the cost of hiring a drill.

Number of days hired → ×6 → +2 → Cost of hiring in pounds

Write the rule in shorthand.
Use *n* to stand for *the number of days hired*
$\quad$ *c* to stand for *the cost of hiring in pounds.*

B2 This machine tells you the cost of hiring a ladder.

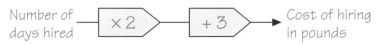

Number of days hired → ×2 → +3 → Cost of hiring in pounds

Write the rule in shorthand.

B3 Write each of these rules in shorthand.

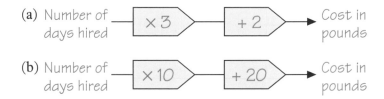

(a) Number of days hired → ×3 → +2 → Cost in pounds

(b) Number of days hired → ×10 → +20 → Cost in pounds

Look at the rule for the cement mixer.

$$n \times 5 + 8 = c$$

Instead of writing $n \times 5$, we usually write $5n$.
$5n$ just means multiply 5 and n together.
So we can write the rule like this.

$$5n + 8 = c$$

(In maths we don't write $n5$. We only write $5n$.)

You can write the rule with c first.

$$c = 5n + 8$$

When you write the rule like this, it is called **a formula**.

B4 Here is a short rule.
Copy and complete this
formula for the rule.

$$n \times 7 + 2 = c$$

$$c = 7n + \ldots\ldots$$

B5 Copy and complete the formulas for each of these rules.

(a)
$$n \times 6 + 3 = c$$
$$c = \ldots n + \ldots$$

(b)
$$n \times 10 - 4 = c$$
$$c = \ldots n - \ldots$$

B6 Write each of these rules as formulas.
(a) $n \times 6 + 10 = c$ (b) $n \times 4 - 2 = c$
(c) $n \times 2 + 25 = c$ (d) $n \times 3 + 12 = c$

B7 This machine gives the price of holidays in Spain.

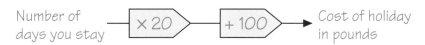

Number of days you stay ——[× 20]——[+ 100]——▶ Cost of holiday in pounds

(a) Let d stand for the *number of days you stay*.
   h stand for the *cost of the holiday in pounds*.

Copy and complete this
short rule for the machine.

$$d \times 20 + \ldots = \ldots$$

(b) Write the rule as a formula.

B8 This machine gives the price of holidays in Greece.
Write the machine **as a formula**.

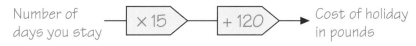

Number of days you stay ——[× 15]——[+ 120]——▶ Cost of holiday in pounds

c Dividing

A farmer rents out a field for car boot sales.
He charges each car for using the field.
The charge depends on how long the car stays in the field.

The rule the farmer uses is

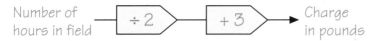

You can write this rule using shorthand.

> h stands for the *number of hours the car is in the field*
> p stands for the *farmer's charge in pounds*.

The rule becomes

$$h \div 2 + 3 = p$$

You can write $h \div 2$ as $\dfrac{h}{2}$

So the formula connecting p and h is,

$$p = \frac{h}{2} + 3$$

C1 Here is a different rule.

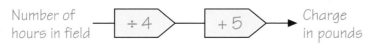

Copy and complete this formula
connecting p and h.

$$p = \frac{h}{\ldots} + \ldots$$

C2 Write each of these rules as formulas connecting p and h.

(a)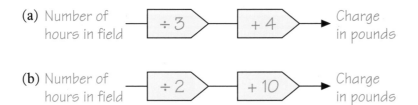

(b)

C3 Write this rule as the formula connecting t and A.

(A is the *Area of field* and t is the *time to clear up*.)

4

D Making formulas

The perimeter of a shape is the distance round its sides.

In this square, each side is 5 cm long. So the perimeter of the square is $5 + 5 + 5 + 5$ cm $= 4 \times 5 = 20$ cm. 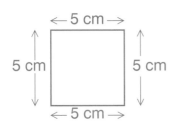	In this square, each side is 8 cm long. So the perimeter of the square is $4 \times 8 = 32$ cm. 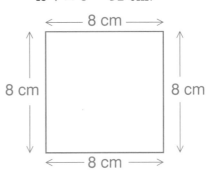

The perimeter of a square is always 4 times the length of one of its sides.

We can write a formula for this.

s stands for the length of the *side of a square in cm*.

p stands for the *perimeter of the square in cm*.

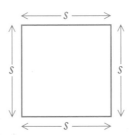

Then the formula connecting *p* and *s* is

$$p = 4s$$

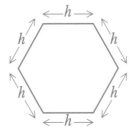

D1 This hexagon has all 6 sides the same length.

(a) Copy and complete
    The perimeter of the hexagon
    is ... times the length of one side.

(b) Let *h* stand for the *length of one side of the hexagon in cm*, and *P* stand for the *perimeter of the hexagon in cm*
    Write down a formula connecting *P* and *h*.

5

D2 This is an equilateral triangle.
All 3 sides are the same length.

Suppose *l* stands for the length
of one side in cm.

P stands for the perimeter of the
triangle in cm.

Write down a formula connecting *P* and *l*.

(Start your formula $P = ...$)

D3 This is a pentagon.
All five sides are the same length.
The length of each side is *s* cm.

The perimeter of the pentagon is *P* cm.

(a) If each side is 4 cm long,
how many cm is the perimeter?

(b) If each side is 100 cm long,
how long is the perimeter?

(c) Copy and complete
The perimeter is ... times the length of one side.

(d) Write down the formula connecting *P* and *s*.

D4 This lorry is carrying crates of apples.
Each crate weighs the same, 20 kg.

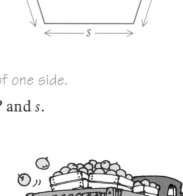

(a) Suppose there are 30 crates on the lorry.
How many kg do 30 crates weigh?

(b) If there are 24 crates on the lorry,
how much do they weigh?

(c) Copy and complete
The weight of the crates = ... × the number of crates.

(d) Suppose *W* stands for the *weight of the crates*
and *n* stands for the *number of crates*.
Write down a formula using *W* and *n*.

E Mixed examples

E1 This is a rule for working out how long it will take
to build a brick wall.

(a) How many days will it take to build a 40 m² brick wall?

(b) Teresa wants a wall built that is 60 m².
How many days will it take to build?

(c) Let A stand for the area of the wall in m².
Let d stand for the number of days it takes to build the wall.
Write down a formula connecting d and A.

E2 A company sells cement in bags.
They charge £5 for each bag of cement and £3 delivery.
You pay £3 delivery whatever the number of bags you buy.

(a) Copy and complete this machine chain for the cost of cement.

(b) Let n stand for the number of bags you buy.
Let T stand for the total cost of the cement, including delivery.
Write down a formula for the cost of buying bags of cement.

E3 The length of each side of this shape is a cm.
The perimeter of the shape is P cm.

(a) How many sides does the shape have?

(b) Write down a formula connecting P and a.

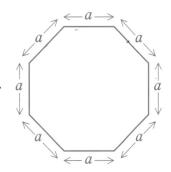

E4 This is the rule for the cost of hiring tables for a party.

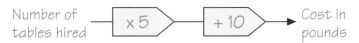

Write this rule as a formula connecting c and N.
(N is the *Number of tables hired*, c is the *cost in pounds*.)

2 Extending coordinates

Central London

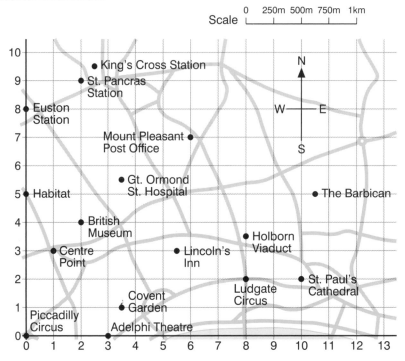

The map shows part of central London.
Piccadilly Circus is at (0, 0).

The British Museum is 2 across and 4 up from Piccadilly Circus.
We say its **coordinates** are **(2, 4)**.
Remember you go **across** then **up**.

> **A1** What has coordinates (1, 3)?
>
> **A2** What is at (10, 2)?
>
> **A3** Write down the coordinates of Mount Pleasant Post Office.
>
> **A4** What are the coordinates of
> (a) St. Pancras Station   (b) Euston Station

Some places have coordinates that are not whole numbers.
For example, King's Cross Station is at (2·5, 9·5)

> **A5** What is at
> (a) (3·5, 1)   (b) (3·5, 5·5)   (c) (5·5, 3)

Each square on the grid is $\frac{1}{4}$ km or 250 m across.

The Adelphi Theatre is at (3, 0).
This is 3 squares across from Piccadilly Circus.

So the distance from Piccadilly Circus to
the Adelphi Theatre is 3×250 m or 750 m.

A6 Euston Station is at (0, 8).
How far is it from there to Habitat?

A7 (a) How far is it from Piccadilly Circus to Euston Station?
(b) About how many minutes do you think it would take
to walk from Piccadilly to Euston Station?

A8 What is 1250 m south of St. Pancras Station?

The map you have been using shows the part of London
which is to the north and to the east of Piccadilly Circus.
This map includes places to the west as well.

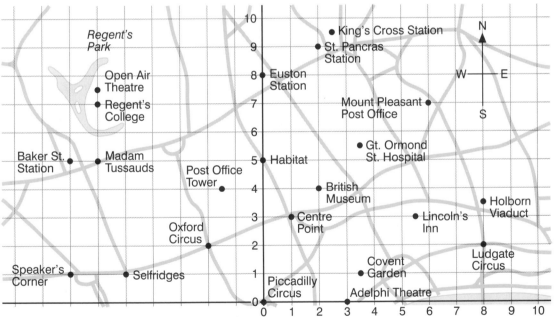

discussion point

Find Regent's College in Regent's Park.
It is 6 squares west of Piccadilly Circus, and 7 squares north.
How could we write its coordinates?
They cannot be (6, 7) because that is Mount Pleasant Post Office.

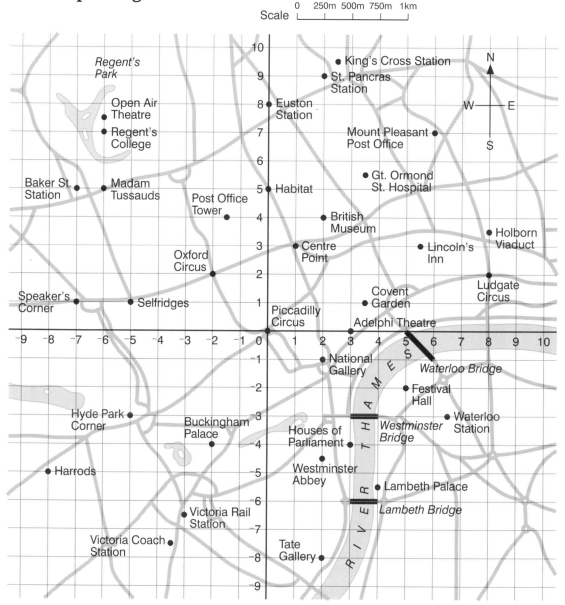

The map above shows a lot more of London.
To write coordinates on this map we need to use negative numbers.

For example, Regent's College is at (⁻6, 7).

> The first coordinate is how far you go **across** from (0, 0).
> If you go to the **left**, the first coordinate is **negative**.

B1 Find Selfridge's, a shop to the west of Piccadilly Circus.
Check that you agree that its coordinates are (⁻5, 1).

B2 What station is at $(^-7, 5)$?

B3 Oxford Circus is north west of Piccadilly Circus.
What are its coordinates?

B4 What is at $(^-1\cdot5, 4)$?

B5 (a) Find the place with coordinates $(^-7, 1)$. What is its name?

(b) How many squares across is it from there to Selfridge's?

(c) How many metres is that?

B6 The Open Air Theatre is 125 m north of Regent's College.
What are the theatre's coordinates?

The second coordinate is how far you go **up** or **down**.
If you go **down**, the second coordinate is **negative**.

For example, the National Gallery is 2 squares across and 1 down from
Piccadilly Circus. Its coordinates are $(2, ^-1)$.

B7 What is at $(3, ^-4)$?

B8 Write down the coordinates of Buckingham Palace.

B9 There is a famous shop at $(^-8, ^-5)$.

(a) What is its name?

(b) From there go 750 m east and 500 m north. Where are you?

B10 Start at $(3, ^-6)$. Cross the bridge. Go 125 m north.
Where are you?

B11 (a) What would you find at $(2, ^-8)$ and at $(2, ^-1)$?

(b) What place is halfway between these two places?

B12 Match each of these chronic clues with the correct coordinates.

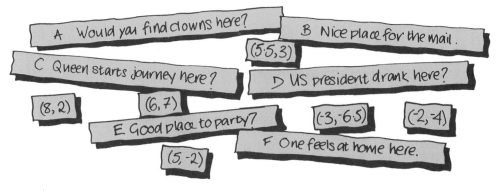

A Would you find clowns here?

B Nice place for the mail.

$(5\cdot5,3)$

C Queen starts journey here?

D US president drank here?

$(8,2)$ $(6,7)$ $(^-3,^-6\cdot5)$ $(^-2,^-4)$

E Good place to party?

F One feels at home here.

$(5,^-2)$

c **Plotting – a game for two players.**

You need a set of number cards numbered from 0 to 5
and another set marked from 0 to ⁻5.
Centimetre squared paper.
Each player needs a different coloured
felt tip.

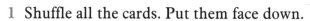

On the paper draw a grid
going from ⁻5 to 5 across and up.
You take turns to mark points on the grid.

1 Shuffle all the cards. Put them face down.

2 The first player takes the top 2 cards.

3 Look at the cards.
You need to plot a point on the grid whose coordinates
are the numbers on the cards.

For example, if the cards are ⁻2 and 3 you could plot
either (⁻2, 3) or (3, ⁻2): you can choose which.

Plot the point and mark it in your colour.
Put the cards back and shuffle the pile.

4 Now the second player takes the top 2 cards and marks their point.

5 Continue taking turns.
The winner is the first person to get 3 points in a straight line.
(Use a ruler to check!)

C1 Copy this diagram
on cm squared paper.
Notice that the across axis
is called the *x-axis*.
The up axis is called the *y-axis*.

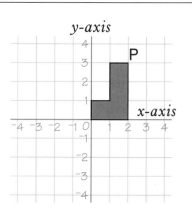

(a) Imagine you stand a mirror
on the *x*-axis.
Draw the reflection of the
shape in the *x*-axis.

(b) Now reflect your reflection
in the *y*-axis.

(c) Reflect the original shape in the *y*-axis.

(d) Write down the coordinates of the point P.

(e) Write down the coordinates of each of the
three reflections of the point P.

12

Review: multiplying

Do NOT use a calculator in this review.
In your answers, show your working clearly.

1   A packet of Killitoff weedkiller weighs 454 grams.
A Garden Centre orders 30 packets of Killitoff.

How many grams of Killitoff are there
altogether in the 30 packets?

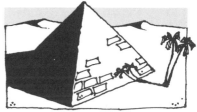

2   A pyramid is made of 631 blocks.
Each block weighs 12 tonnes.
How much does the pyramid weigh altogether?

3   A Greek theatre was divided into 16 sections.
Each section held 435 people.
How many people did the theatre hold altogether?

4   Some diabetics have to have insulin injections.
Alice has had an injection of insulin every day for 32 years.
How many injections is this?
(Use 1 year = 365 days.)

5   Selma walks 22 km to school each week.
She goes to school 38 weeks a year.
How far does she walk in a year?

6   Ira buys two magazines a week. They cost £1·49 and £1·75 each.
How much does she spend each year on magazines?
(Remember: a year is 52 weeks.)

7   Alan is looking for a job.
How much a year would
he get in each of these jobs?

3 Making shapes

A Pyramids and prisms

When you hear
the word **pyramid**
what do you think of?

Most people would
think of the
pyramids in Egypt.

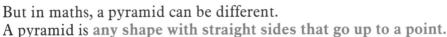

But in maths, a pyramid can be different.
A pyramid is **any shape with straight sides that go up to a point.**

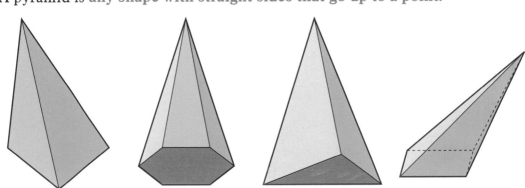

A1 Which of these are pyramids?
(Remember: **straight** sides up to a **point.**)

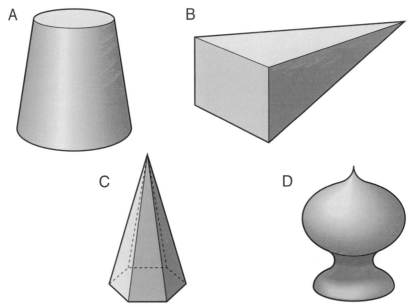

14

When they hear the word **prism**
most people think of something like this.

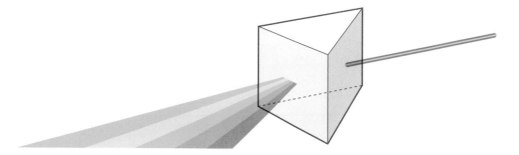

But in maths, a prism is any shape made up of **slices exactly the same
all the way through**.

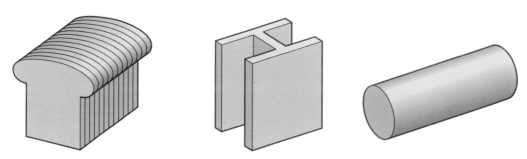

A2   Which of these shapes are prisms?
      (Remember: you must be able to cut the shape into slices
      that are the **same shape and size** all the way through.)

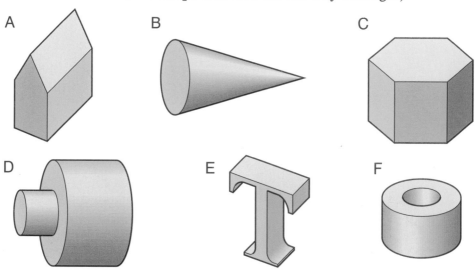

A   B   C

D   E   F

A3 Some of these shapes are pyramids.
Some are prisms. Some are neither.
For each one, write Pyramid, Prism or Neither.

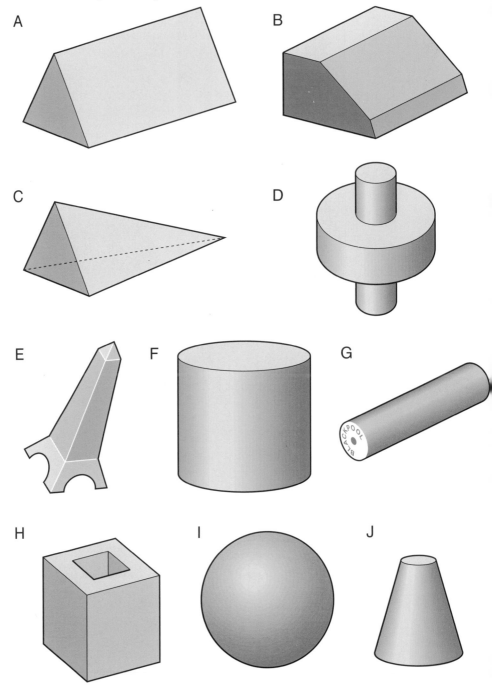

A

B

C

D

E

F

G

H

I

J

If you cut up a shape and open it out you get its **net**.

A4 Which of these shapes could be the net of this cuboid?

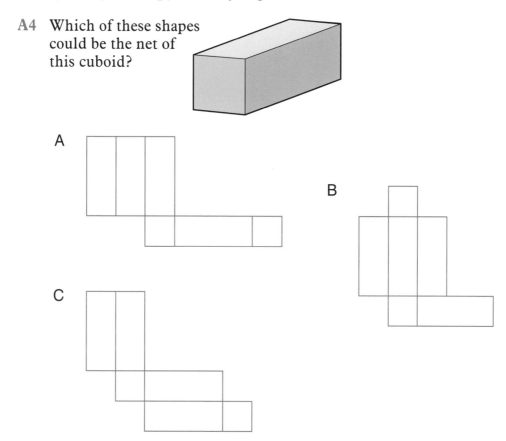

A

B

C

A5 This is the net for a pyramid.

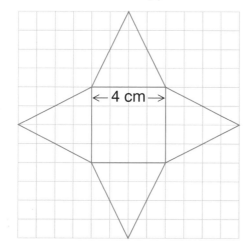

←— 4 cm —→

Draw the net accurately on squared paper.
Add tabs to every other edge of the net.
Then cut the net out and make the pyramid.

Try it!

An experiment for three people.

Here is the net of another pyramid.

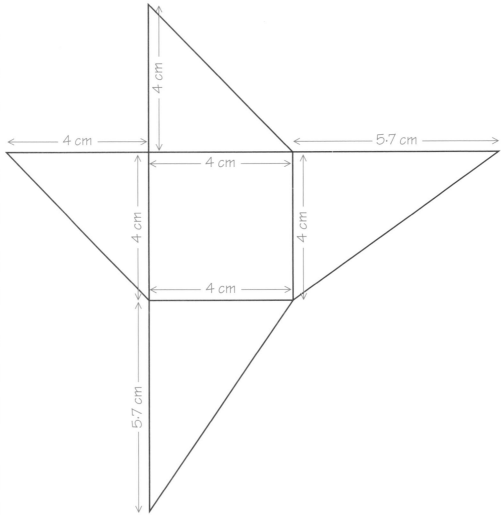

Each person needs to draw the net carefully on squared paper.
Add tabs to every other edge.
Fold and make up your three pyramids.

You can fit the three pyramids together to make a cube.
Can you see how?

Try making other shapes with more of the same pyramids.

B Drawing triangles

To draw some nets you need to be able to draw
triangles accurately.

Here is a rough sketch of a triangle.
It is not drawn at all accurately,
but it shows how long the sides are.

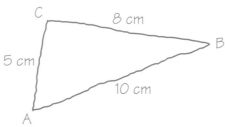

Get a ruler and pencil.
See if you can draw the triangle
accurately by trial and error, like this.

1 Draw the line from A to B first.
Make AB exactly 10 cm long.

A   10 cm   B

2 Now guess where C must be
and draw a point.
Measure CB and CA.
Were you far out?

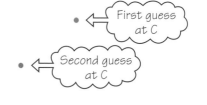

3 Have another guess at C.
Measure CB and CA.
Keep going until you get
C in exactly the right place.

A   B

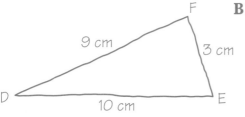

B1 Draw this triangle by the method above.

Then check how accurate you were
by measuring each angle.
The angle at D should be about 17°,
E about 62°, and F about 101°.

19

You need compasses, ruler and pencil.

Drawing a triangle by trial and error
can waste a lot of time.

Here is a more accurate way.

Follow these instructions
to draw this triangle accurately.

**Not
to scale**

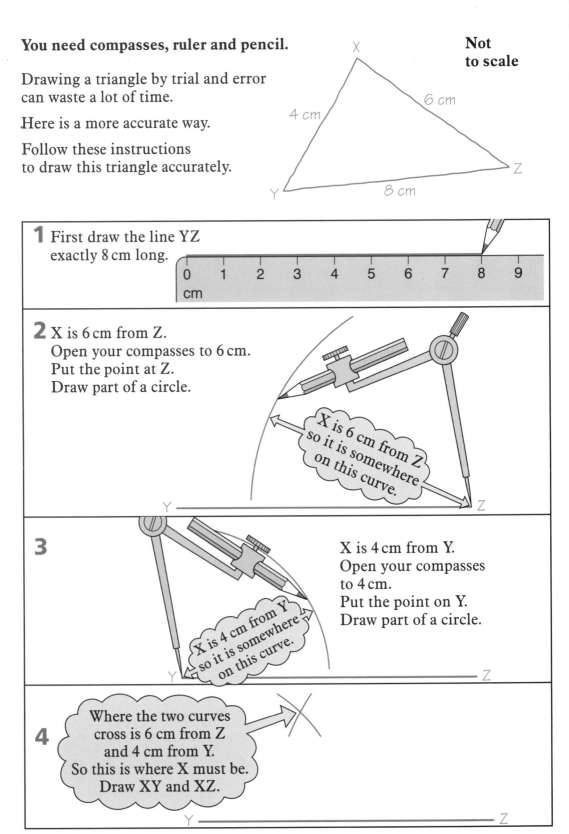

X

4 cm

6 cm

Y

8 cm

Z

1 First draw the line YZ
exactly 8 cm long.

0   1   2   3   4   5   6   7   8   9
cm

2 X is 6 cm from Z.
Open your compasses to 6 cm.
Put the point at Z.
Draw part of a circle.

X is 6 cm from Z
so it is somewhere
on this curve.

Y   Z

3

X is 4 cm from Y
so it is somewhere
on this curve.

X is 4 cm from Y.
Open your compasses
to 4 cm.
Put the point on Y.
Draw part of a circle.

Y   Z

4

Where the two curves
cross is 6 cm from Z
and 4 cm from Y.
So this is where X must be.
Draw XY and XZ.

Y   Z

B2 Here are sketches of two triangles.
Draw each triangle accurately, using compasses and ruler.

(a)   (b)

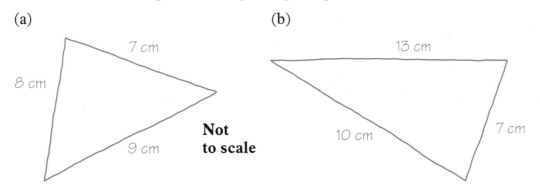

B3 The sketch below shows the net of a pyramid.
Draw the net accurately.
Start by drawing the base on squared paper.
Then draw the triangles round the base.

You **must use compasses and ruler** to draw the triangles.
The squared paper will not help.

Check your drawing by making the pyramid.
(Add flaps to every other edge before cutting it out.)

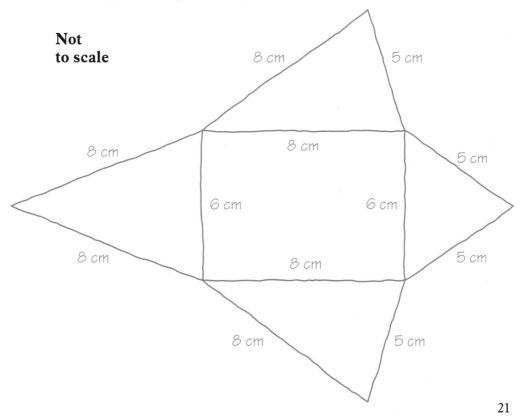

C Nets

C1 This is a small Toblerone box.
The box is a prism.

This is a sketch of part of the net of the Toblerone box.

Draw the complete net on squared paper.
You will have to use compasses and a ruler to draw the triangles.

Add tabs to your net.
Then cut it out and make the box.

22

C2 A sales company is experimenting with a design for a new box.
This is a sketch showing the dimensions of the new box.

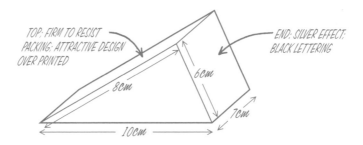

Here is a sketch of the net for the box.
Not all the dimensions are on the sketch.

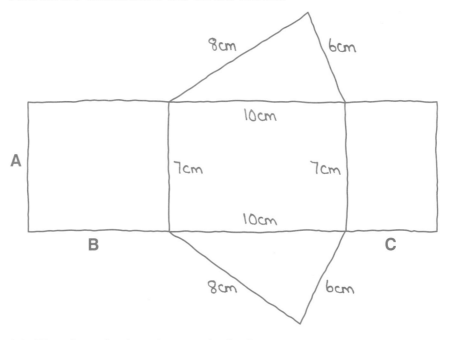

(a) How long is the edge marked **A**?

(b) How long is the edge marked **B**?
(Hint: look at where this edge is in the sketch of the box.)

(c) How long is edge **C**?

(d) Draw the net accurately on squared paper.
Start by drawing the base, 10 cm by 7 cm.
Then draw the end and top.
Last, use compasses and ruler to draw
the triangular sides.

This is a square
based pyramid.

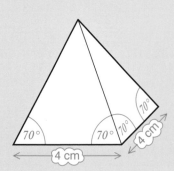

And here is a sketch of its net.

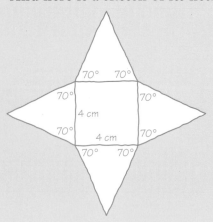

You need an angle measurer.
Here is how to draw the net.

1 First draw a 4 cm square.

4 cm

4 cm

2 With an angle measurer,
draw an accurate angle of 70°.

70°

Then lightly
draw this
edge.

3

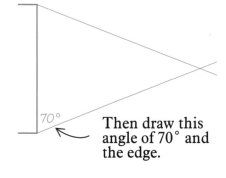

70°

Then draw this
angle of 70° and
the edge.

4 Draw in the edges you want
more heavily.

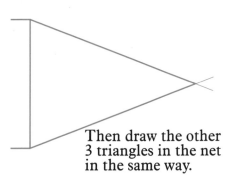

Then draw the other
3 triangles in the net
in the same way.

C3 Draw the net of the pyramid on the opposite page.

C4 Here is another pyramid.
The base is a square
with sides 5 cm.
The faces have angles of
55° at the bottom.

Draw an accurate
net for the pyramid.

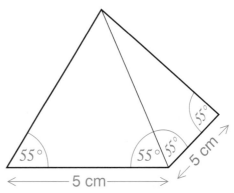

C5 Here is a sketch of a net.

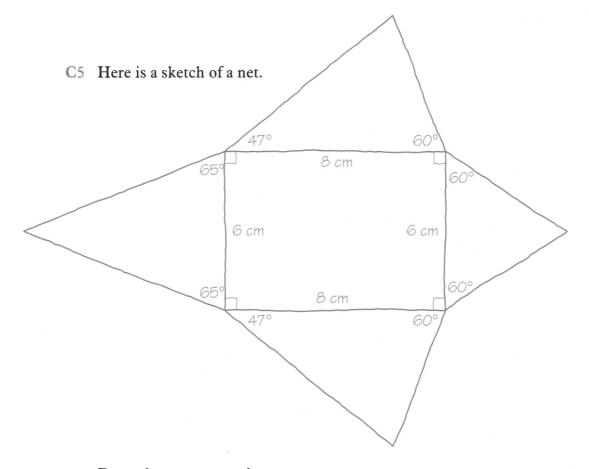

Draw the net accurately.
(Start with the rectangle.)
Add tabs to your net and make the solid.

Review: flowcharts

1 Draw two columns.
Label them A and B.
Write 1 at the top of column A.
Write 1 at the top of column B as well.

Now follow the instructions in this flow chart.

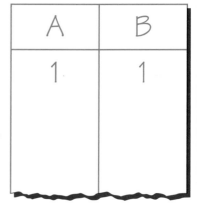

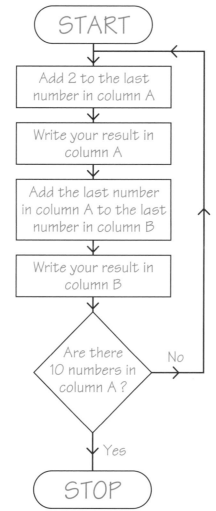

(a) What special name do we give to the numbers in column A?

(b) What special name do the numbers in column B have?

3 Look at the flow chart below.
You have to start with a three digit number.
A number with three digits is any number like

287 or 381 or 910 or 742 or 663 or ...

On the right there is an example started for you.

Work through the flow chart. Start with 451.
Write down the numbers the chart asks you to.

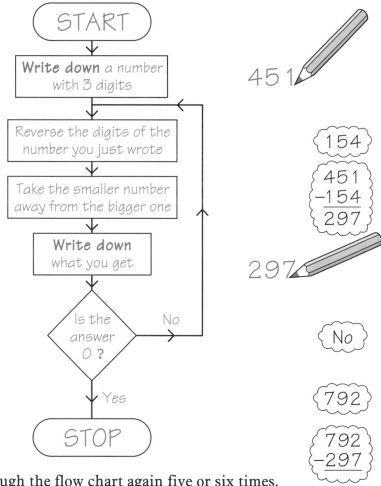

START

Write down a number
with 3 digits

451

Reverse the digits of the
number you just wrote

154

Take the smaller number
away from the bigger one

451
−154
297

Write down
what you get

297

Is the
answer
0 ?   No

No

Yes

792

STOP

792
−297

(a) Go through the flow chart again five or six times.
Start with a different number each time.

(b) Do you always end up at 0?

(c) What number do you get before you get 0?

(d) Do you always get the same number before that?
If not, what numbers can you get?

(e) Look at each of the sets of numbers you got from the flow chart.
Write about anything similar you can see in them.

27

4 Old and new units

A Weights and measures

There are still lots of places where we don't use metric units.

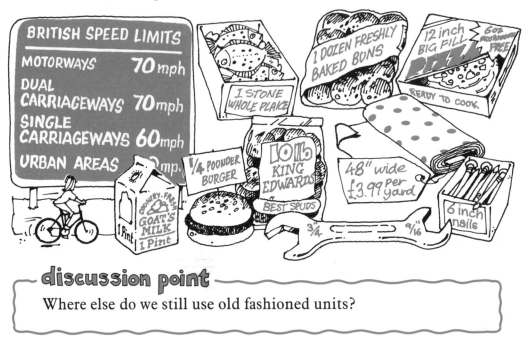

> ### discussion point
>
> Where else do we still use old fashioned units?

Old units like pounds and ounces, or pints can be difficult to use.
But you can easily convert them to metric units.

1 pound (1 lb) is about half a kilogram.

1 ounce (1 oz) is about 25 grams.

1 pint (1 pt) is about half a litre.

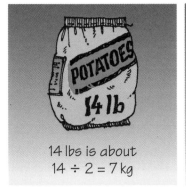

14 lbs is about
14 ÷ 2 = 7 kg

4 oz is about
4 × 25 = 100 g

6 pts is about
6 ÷ 2 = 3 litres

28

A1 Roughly what do each of these weigh in metric units.

A2 About how many litres are in each of these?

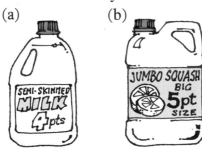

A3 Sandra is on holiday in France.
Here is a recipe she wants to cook.
Write out a shopping list for her using metric units.

She does not mind having too much of something.
But she doesn't want less!

BEEF IN RED WINE
Serves 12
6 pts red wine or stock
4 lb stewing beef
2 lb carrots
3 lb onions
1 lb turnips
4 oz. flour
Fry the beef in the casserole

A4 This is the cost of the ingredients in Francs.

Red wine – 5F a litre. Beef – 25F a kilo. Carrots – 12F/kilo
Onions – 8F/kilo. Turnips – 10F/kilo. Flour – 1F per 100g

(a) Work out roughly how much the Beef in Red Wine costs in Francs.

(b) 8 Francs are about £1. What does her stew cost in pounds?

B Distances

In Britain we use miles to measure long distances.
Many other countries use kilometres (km).
You can change from km to miles quite easily.

	Le Mans to Paris is 200 km.
Halve the number of kilometres	$\frac{1}{2}$ of 200 = 100
Find a quarter of what you get.	$\frac{1}{4}$ of 100 = 25
Add it on.	125
That's the distance in miles.	Le Mans to Paris is 125 miles.

B1 Work out each of these distances in miles.

(a) 40 km   (b) 60 km   (c) 6 km

B2 From Paris to Monte Carlo is 1000 km.

(a) How many miles is that?

(b) About how long would it take to drive
at an average speed of 60 m.p.h.?

B3 Sandra gets a letter from her French pen-friend.
Write out the letter again.
Put all the distances in miles.

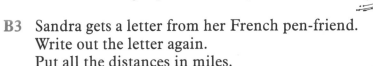

Dear Sandra,
Our new house is really nice – but it is 20 km to the nearest town! I have to cycle 8 km to work each day (better than Étienne – he has to go 16 km to work!). At least Paris is only 40 km by rail, much closer than when we were 90 km away.

B4 Here are some world records. Change them into miles.

(a) In 1988, a Colombian cycled 58 km in one hour.

(b) In 1991, a Mexican ran 21 km in one hour.

(c) In 1988, a Dutchman skated 546 km in one day.

(d) In 1941, a pigeon flew 1292 km in one day.

c Do It Yourself

Most things in DIY shops are metric.
But many things in the home are not.

It is very easy to convert inches to metric measurements.

1 inch is $2\frac{1}{2}$ cm (almost exactly)

Suppose you want to convert 6 inches to centimetres.
You need to multiply 6 by $2\frac{1}{2}$
You can do this in lots of ways.

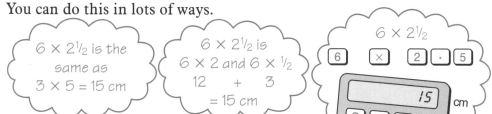

C1 Convert each of these measurements into centimetres.

(a)

8 INCH FLAN BASE

(b) ←—10 inches—→ 12 inches

(c) 3 inches

C2 Suni wants to buy some computer labels.
The labels are 4 inches wide and 2 inches deep.

What are the dimensions of the labels in centimetres?

Tamás want to buy a woodscrew $1\frac{1}{2}$ inches long.
In the DIY shop all woodscrews are measured in mm.

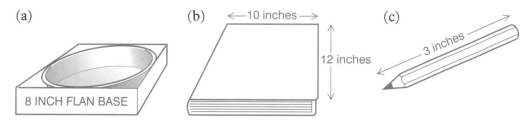

C3 How long are each of these woodscrews in mm?

(a) $2\frac{1}{2}$ inches

(b) $3/4$ inch

(c) $1\frac{1}{4}$ inches

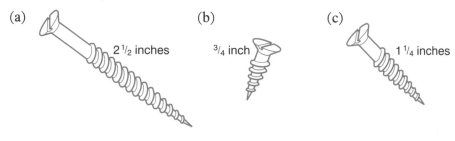

D Biggest and smallest

Here are some big and small measurements in old units.
Convert each of them into metric units.

An ostrich can be up to 96.
inches high and weigh over
300 lb.

The largest champagne bottle is
called a Nebuchadnezzar. It
holds 26 pints.

In the 1890s, Joe Darby could
jump over a full-size snooker
table 150 inches long. He landed
on a crate of eggs – without
breaking any!

The wandering albatross has a
wingspan of 120 inches!

A Great Dane can be over 36
inches at the shoulder and weigh
more than 180 lb.

Limbo dancers can dance under
a bar less than 8 inches off the
ground.

The chihuahua can be less than
6 inches high and weigh less
than 2 lb.

Review: dividing

Do NOT use a calculator in this review.
In your answers, show your working clearly.

1 37 students hire a coach to go to a concert.
   The coach costs £259 to hire.
   How much should each student pay?

2 Trevor is packing Easter eggs into boxes.
   He has to pack 315 eggs into 35 boxes.
   How many eggs should go into each box?

3 Bob is a terrible speller.
   He writes a story 18 pages long.
   It has 306 spelling mistakes in it!
   How many mistakes is that on
   each page on average?

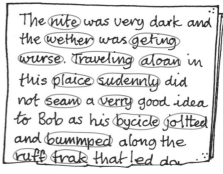

4 A train is carrying 384 passengers. It breaks down.
   The passengers have to get onto coaches to carry on.
   Each coach holds 48 passengers.
   How many coaches are needed?

5 Susan is packing eggs onto trays.
   Each tray holds 36 eggs.
   She has 864 eggs to pack.
   How many trays does she need?

6 The largest beetroot ever grown weighed 464 ounces.
   How many pounds did the beetroot weigh?
   (There are 16 ounces in a pound.)

7 In America people know their weights in pounds.
   Bill weighs 210 pounds. How many stones is this?
   (14 pounds make 1 stone.)

8 Janice is packing eggs into boxes. Each box contains 12 eggs.
   She has to pack 500 eggs. How many full boxes will these fill?
   How many eggs are left over?

33

5 Negative numbers 1

A Adding

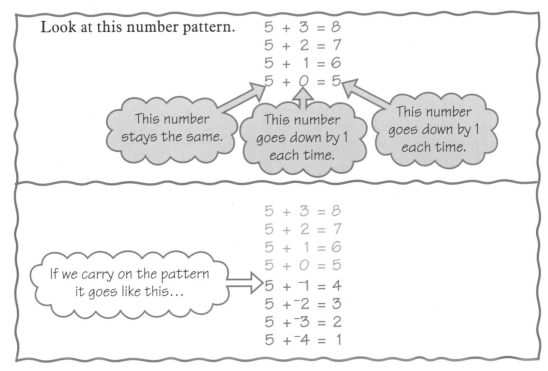

Look at this number pattern.

$$5 + 3 = 8$$
$$5 + 2 = 7$$
$$5 + 1 = 6$$
$$5 + 0 = 5$$

This number stays the same.

This number goes down by 1 each time.

This number goes down by 1 each time.

$$5 + 3 = 8$$
$$5 + 2 = 7$$
$$5 + 1 = 6$$
$$5 + 0 = 5$$
$$5 + {}^-1 = 4$$
$$5 + {}^-2 = 3$$
$$5 + {}^-3 = 2$$
$$5 + {}^-4 = 1$$

If we carry on the pattern it goes like this...

A1 Copy each of these patterns. Fill in the missing numbers.

(a)
$$9 + 4 = 13$$
$$9 + 3 = 12$$
$$9 + 2 = 11$$
$$9 + 1 = 10$$
$$9 + 0 = 9$$
$$9 + {}^-1 = 8$$
$$9 + {}^-2 = 7$$
$$9 + {}^-3 = ...$$
$$9 + {}^-4 = ...$$

(b)
$$6 + 3 = 9$$
$$6 + 2 = 8$$
$$6 + 1 = 7$$
$$6 + 0 = 6$$
$$6 + {}^-1 = 5$$
$$6 + {}^-2 = 4$$
$$6 + {}^-3 = ...$$
$$6 + ... = ...$$
$$6 + ... = ...$$

(c)
$$8 + 4 = 12$$
$$8 + 3 = 11$$
$$8 + 2 = 10$$
$$8 + 1 = 9$$
$$8 + 0 = 8$$
$$8 + {}^-1 = 7$$
$$8 + {}^-2 = ...$$
$$... + ... = ...$$
$$... + ... = ...$$

A2 Look carefully at each of these addition patterns.
Then write down the next three lines.

(a)
$$7 + 4 = 11$$
$$7 + 3 = 10$$
$$7 + 2 = 9$$
$$7 + 1 = 8$$
$$7 + 0 = 7$$
$$7 + {}^-1 = 6$$

(b)
$$10 + 3 = 13$$
$$10 + 2 = 12$$
$$10 + 1 = 11$$
$$10 + 0 = 10$$
$$10 + {}^-1 = 9$$
$$10 + {}^-2 = 8$$

(c)
$$13 + 2 = 15$$
$$13 + 1 = 14$$
$$13 + 0 = 13$$
$$13 + {}^-1 = 12$$
$$13 + {}^-2 = 11$$
$$13 + {}^-3 = 10$$

Look at the patterns

You can see $3 + {}^-1 = 2$. But $3 - 1 = 2$ as well.	And $5 + {}^-2 = 3$. But $5 - 2 = 3$ as well.

> **If you add $^-1$, it is the same as subtracting 1.**
> **If you add $^-2$, it is the same as subtracting 2.**

You can use this rule when adding any negative numbers.

For example, to work out $4 + {}^-3$,
you would write

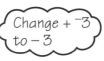

$$4 + {}^-3$$
$$= 4 - 3$$
$$= 1$$

Change + $^-3$ to $- 3$

A3 Work out each of these.
Write them out like the example above.
(a) $6 + {}^-2$ (b) $5 + {}^-3$ (c) $3 + {}^-1$
(d) $12 + {}^-8$ (e) $5 + {}^-4$ (f) $4 + {}^-4$
(g) $8 + {}^-3$ (h) $10 + {}^-7$ (i) $4 + {}^-2$
(j) $6 + {}^-5$ (k) $9 + {}^-5$ (l) $8 + {}^-6$

Sometimes your answer may be a negative number.

For example, to work out
$4 + {}^-6$ you would write
$$4 + {}^-6$$
$$= 4 - 6$$
$$= {}^-2$$

A4 Work out each of these,
(a) $2 + {}^-5$ (b) $1 + {}^-3$ (c) $5 + {}^-10$
(d) $1 + {}^-8$ (e) $2 + {}^-4$ (f) $3 + {}^-7$
(g) $8 + {}^-13$ (h) $1 + {}^-7$ (i) $4 + {}^-12$

A5 Work out these in the same way. Use a number line.
Check with your teacher after you have done (a).
(a) $^-2 + {}^-5$ (b) $^-1 + {}^-4$ (c) $^-5 + {}^-5$
(d) $^-1 + {}^-1$ (e) $^-2 + {}^-4$ (f) $^-3 + {}^-7$
(g) $^-8 + {}^-3$ (h) $^-1 + {}^-7$ (i) $^-4 + {}^-8$

B Subtracting

Here is a different pattern of numbers.
This is a subtraction pattern.

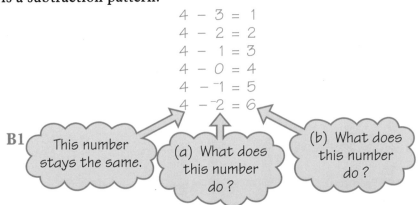

$4 - 3 = 1$
$4 - 2 = 2$
$4 - 1 = 3$
$4 - 0 = 4$
$4 - {}^-1 = 5$
$4 - {}^-2 = 6$

B1 This number stays the same.

(a) What does this number do?

(b) What does this number do?

B2 (a) Copy the pattern and write down the next three lines.

(b) Use the pattern to find the answer to $4 - {}^-5$.

B3 Copy these patterns and fill in the missing numbers.

(a)
$3 - 1 = 2$
$3 - 0 = 3$
$3 - {}^-1 = 4$
$3 - {}^-2 = 5$
$3 - {}^-3 = \dots$
$3 - {}^-4 = \dots$
$3 - {}^-5 = \dots$

(b)
$6 - 2 = 4$
$6 - 1 = 5$
$6 - 0 = 6$
$6 - {}^-1 = 7$
$6 - {}^-2 = \dots$
$6 - \dots = \dots$
$\dots - \dots = \dots$

B4 Use the patterns in question B3 to work out

(a) $3 - {}^-4$ (b) $3 - {}^-5$ (c) $6 - {}^-2$

| You can see | $3 - {}^-2 = 5$ | And | $6 - {}^-4 = 10$ | |
| But | $3 + 2 = 5$ | But | $6 + 4 = 10$ | as well. |

> Subtracting $^-1$ is the same as adding 1.
> Subtracting $^-2$ is the same as adding 2.

You can use this rule when subtracting any negative numbers.

For example, to work out $6 - {}^-3$,

$6 - {}^-3$
$= 6 + 3$
$= \underline{9}$

Change $- {}^-3$ to $+ 3$

36

B5 Work out each of these.
Write them out like the example at the bottom of page 36.
- (a) 6 – ⁻4
- (b) 6 – ⁻1
- (c) 3 – ⁻1
- (d) 10 – ⁻8
- (e) 2 – ⁻1
- (f) 4 – ⁻4
- (g) 2 – ⁻3
- (h) 0 – ⁻6
- (i) 1 – ⁻9

You can easily work out ⁻3 – ⁻7.

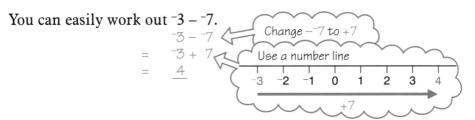

$$⁻3 – ⁻7$$
$$= \quad ⁻3 + 7$$
$$= \quad \underline{4}$$

Change –⁻7 to +7

Use a number line

-3 -2 -1 0 1 2 3 4

+7

B6 Work out these. Check with your teacher after (a).
- (a) ⁻2 – ⁻6
- (b) ⁻3 – ⁻8
- (c) ⁻5 – ⁻1
- (d) ⁻7 – ⁻2
- (e) ⁻4 – ⁻4
- (f) ⁻10 – ⁻3

B7 Work out each of these.
Be careful! Some are subtractions and some are additions.
- (a) ⁻6 – ⁻5
- (b) 8 + ⁻4
- (c) 6 + ⁻10
- (d) ⁻10 + 8
- (e) ⁻2 + ⁻7
- (f) ⁻4 – ⁻14
- (g) ⁻2 – ⁻8
- (h) 0 + ⁻9
- (i) ⁻1 + ⁻9

B8 There is a formula that tells you the difference in temperature between the inside and outside of a house.

The formula is $d = a - b$.
d stands for the difference in temperature in degrees Celsius.
a stands for the inside temperature.
b stands for the outside temperature.

For example, when $a = 20$ and $b = ⁻2$, then
$$d = 20 – ⁻2$$
$$d = 20 + 2$$
$$d = \underline{22}$$

Work out the value of d when
- (a) $a = 21$ and $b = ⁻5$
- (b) $a = 20$ and $b = ⁻8$
- (c) $a = 15$ and $b = ⁻4$
- (d) $a = 19$ and $b = ⁻7$

6 Symmetry in 3D

In this chapter you will be using multilink cubes.
You should ignore the holes and connectors.

A Reflection symmetry

1 Make two of these shapes with multilink cubes.

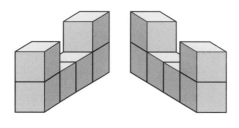

2 Put the shapes side by side. This makes a bigger shape.

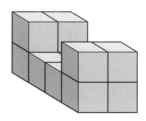

3 Now put a mirror in the middle of the bigger shape like this.

Take the mirror away, then put it back.
The bigger shape does not change.
One side is the reflection of the other.

We say the shape has **reflection symmetry**.
The mirror is a **plane of symmetry**.

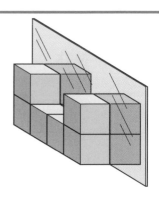

A1 Does the bigger shape have any other planes of symmetry?

Check with a mirror.
(You will need to split the shape in two.)

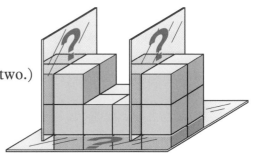

A2 Make this shape from 16 multilink cubes. Does it have any planes of symmetry?

Check your answer with a friend.
(Hint: there is more than one plane.)

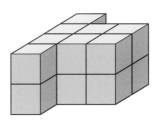

A3 Here are some everyday objects.
Which of them have planes of symmetry?

For the ones that do have planes of symmetry, say how many the object has.

(a)　　　　　　　　　(b)　　　　　　　　　(c)

(d)　　　　　　　　　(f)　　　(g)

(e)

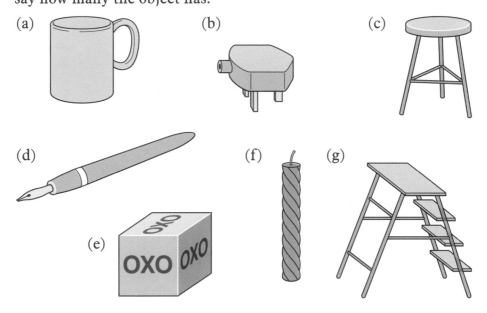

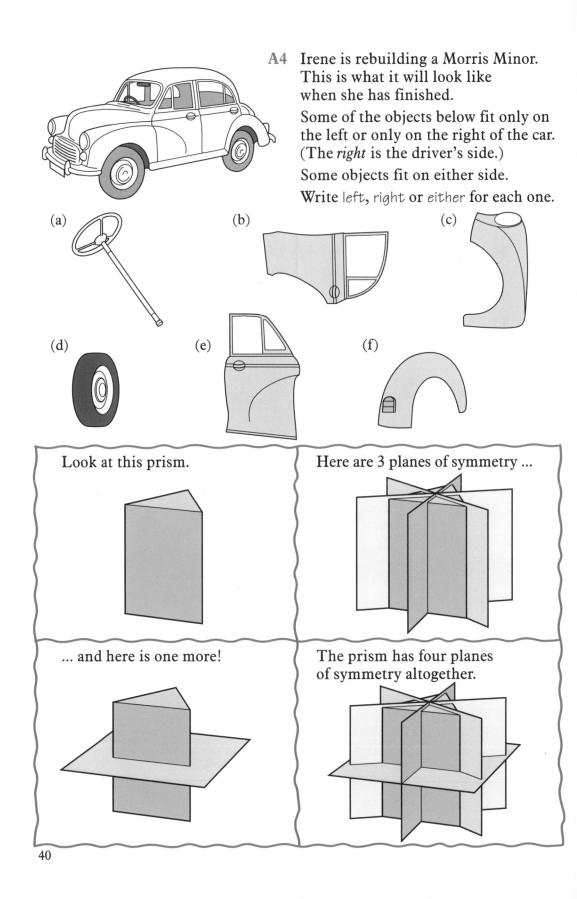

A4 Irene is rebuilding a Morris Minor. This is what it will look like when she has finished.

Some of the objects below fit only on the left or only on the right of the car. (The *right* is the driver's side.)

Some objects fit on either side.

Write left, right or either for each one.

(a)

(b)

(c)

(d)

(e)

(f)

Look at this prism.

Here are 3 planes of symmetry ...

... and here is one more!

The prism has four planes of symmetry altogether.

A5 Here are some shapes.
How many planes of symmetry does each of them have?
You may be able to build some of them
with multilink.

(a)

(b)

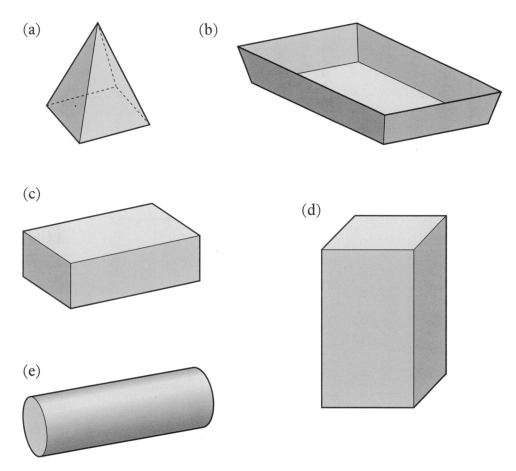

(c)

(d)

(e)

A6 Make a shape using 5 multilink cubes
that has no planes of symmetry.

A7 Make shapes from 5 multilink cubes that have

(a) only 1 plane of symmetry

(b) only 2 planes of symmetry

B Rotation symmetry

Make this shape from multilink cubes.
Put the cubes together so that a pencil
will fit down the middle.

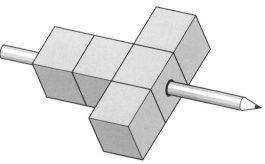

Rotate the shape round the pencil.
Check that there are two positions where it looks the same.

The line of the pencil is called an **axis of rotation symmetry**.

The solid shape looks the same in 2 positions.
We say it has **rotation symmetry of order 2**
about the pencil.

B1 Add one more cube to your shape to make this shape.

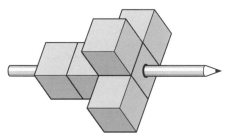

Does the shape have rotation symmetry now?

B2 Add another cube to make this shape.

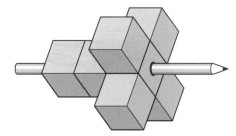

(a) Rotate the shape round the pencil.
In how many positions does it look the same?

(b) What is the order of rotation symmetry about the pencil?

B3 Add 4 more cubes to make this shape.

(a) What is the order of rotation symmetry about the pencil now?

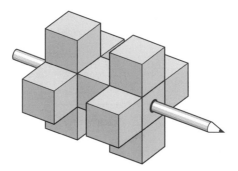

(b) Imagine that you could stick a pencil through the shape like this.

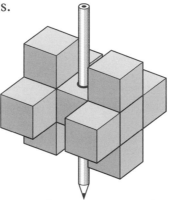

The shape also has rotation symmetry about this new axis. What is its order of rotation symmetry about this axis?

B4 (a) What is the order of rotation symmetry of this shape about pencil A?

(b) What is the order of rotation symmetry of the shape about pencil B?

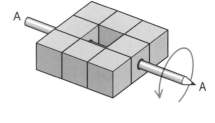

(c) Could you imagine a pencil through the shape anywhere else so that it has rotation symmetry about that pencil?

Try to sketch the shape and show where you could put the pencil.

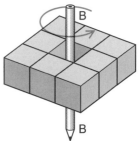

43

B5 Each of these shapes has rotation symmetry
about the axis shown.
What is the order of rotation symmetry of each shape
about the axis?

(a)

(b)

(c)

(d)

(e)

(f)

(g)

(h)

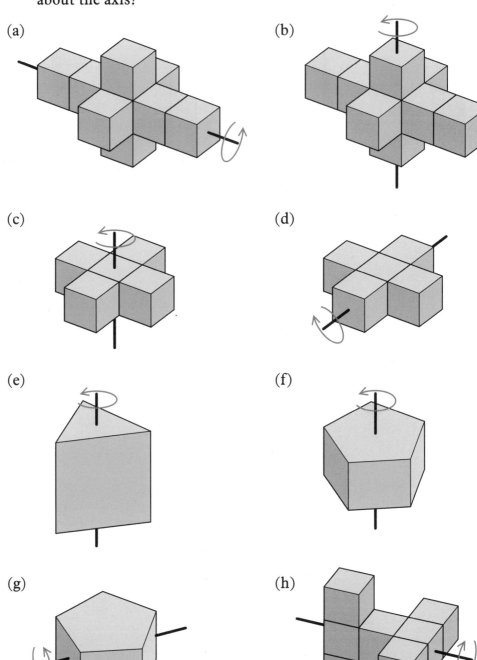

B6 Some of these shapes have rotation symmetry.
Some do not.

For each shape, say whether it has rotation symmetry.
Discuss your answers with a friend.

(a)

(b)

(c)

(d)

(e)

(f)

(g)

(h)

(i)

(j)

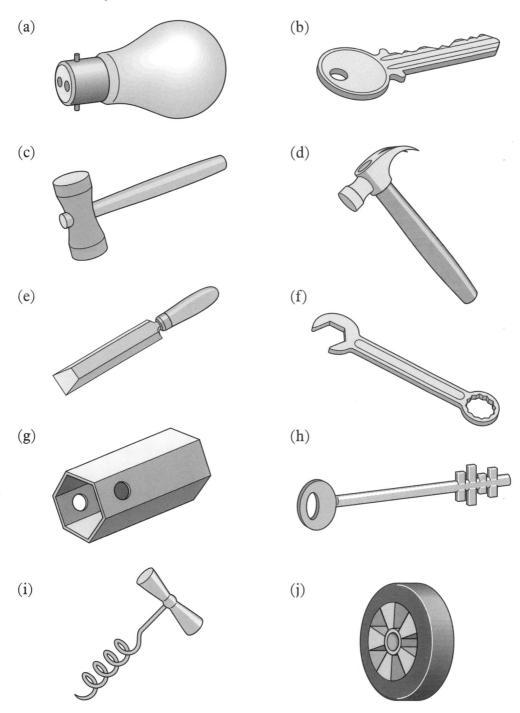

Discussion: Surveys

When you carry out a survey, you must plan carefully.

You may ask people questions.
You have to be careful who you ask,
and what you ask them.

How popular is
vegetarian food?

You may be noting down things.
You have to be careful how
you go about it.

What types of shoes
do people wear?

Here are some surveys.
There is quite a lot wrong with them!

Look at each survey.
Talk about what things are wrong with it.
How could you improve it?

The *Big Porridge Company* wanted
to find out what people ate for
breakfast.

They picked out a page at random
from the phone book.
Then they phoned the first 100 people
on the page.

They asked them what they had for
breakfast that morning.

Over 40% of the people had porridge.

2 Tracey wanted to find out how pupils come to school.
She prepared this table to record her results.

Method	Tally	Total
Bus		
Bicycle		
Walk		

She got to school an hour before it started.
She asked the first 50 pupils "How did you get to school today?"

3 *Media Mags* wanted to find out what magazines people read.
They picked out 250 names from different pages of the phone book.
Then they phoned up between 2pm and 5pm.
They asked the person who answered the phone
"What magazines have you read this week?"

Type of magazine	% age
	18·3
TV magazines	12·7
Puzzle magazines	6·5

4 Sean wanted to find how much homework pupils at his school did.
He asked all the pupils in his class "How much homework do you usually do?"
He kept a note of the replies.
At the end he drew a bar chart to show the results.

1. Usually 15 mins, but last night about $\frac{1}{2}$ an hour.
2. Always do it at school.
3. It varies between 20 minutes and an hour.
4. Didn't do any last night.
5. 2 hours at weekends, but less in the week.
6. Last night about 20 minutes, on Tuesdays
   about 40 minutes and on Monday I did about

7 Negative numbers 2

A Multiplying

A1 Look at this number pattern (we call it a **sequence**).

16, 12, 8, 4, 0, ⁻4, ⁻8, ..., ..., ...

Write down the next three numbers in the sequence.

A2 Continue these sequences.

(a) 6, 4, 2, 0, ⁻2, ..., ..., ...

(b) 4, 3, 2, 1, 0, ..., ..., ...

(c) 6, 3, 0, ..., ..., ...

(d) 20, 15, 10, 5, ..., ..., ...

A3 **You need worksheet G9–2.**

Look at column (a) on the worksheet.
The numbers in the sequence go down in 1s.
Complete the column.

A4 Complete columns (b), (c) and (d) by continuing the sequences.

Look at the sequence of numbers in the top row of the worksheet.

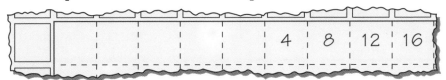

As you go across the row, from right to left,
the numbers go down in fours.

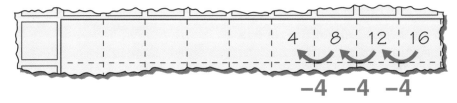

A5 Complete the numbers across the whole of the top row.

A6 Look at the numbers in the second row.

(a) What do they go down in, as they go from right to left?

(b) Complete the second row.

(c) Complete the third and fourth rows.

48

A7 The fifth row should be very easy to complete!
Complete the fifth row and check your table with a friend.

Check that the next row in your table has these numbers in it.

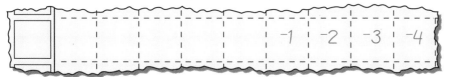

As you go from right to left, the numbers get bigger by 1.

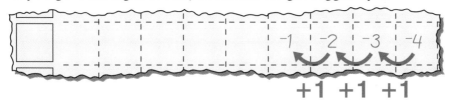

$$+1 \quad +1 \quad +1$$

A8 Complete the numbers in this row.

A9 Look at the next row.
These numbers also get bigger as you go from right to left.
(a) Complete this row.
(b) Complete the last two rows in the table.

Now check your table with your teacher.

A10 Copy and complete this multiplication square.

X	1	2	3	4
4	4	8	12	
3	3			
2			6	
1				

2 × 3

Look at your multiplication square.
It is the same as the top right-hand part
of the table on the worksheet.

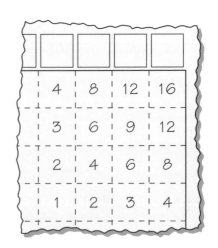

A11 On the worksheet, put the numbers 1, 2, 3, 4 into these boxes.

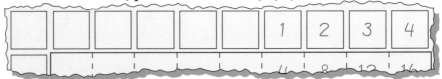

A12 Look at the numbers in the boxes at the top of the worksheet.
The numbers form a pattern as you go from right to left.

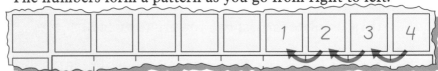

Use the pattern to fill in the rest of the boxes
at the top of the worksheet.

A13 Write the numbers 4, 3, 2, 1
in the boxes at the left side
of the worksheet.

They form a pattern
as you go down.

Use the pattern to fill in
the rest of the boxes.

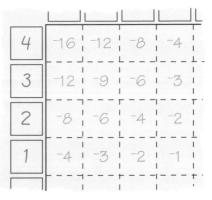

A14 Now you have a giant multiplication square.
You can use it to multiply negative numbers.
Use your table to check that these calculations are correct.
(a) $4 \times {}^-2 = {}^-8$ (b) $4 \times {}^-1 = {}^-4$
(c) ${}^-3 \times {}^-2 = 6$ (d) ${}^-3 \times 4 = {}^-12$

50

A15 Use your table to work out these.

(a) $^-4 \times {}^-3$ (b) $^-2 \times 3$ (c) $3 \times {}^-4$

(d) $^-1 \times 4$ (e) $1 \times {}^-3$ (f) $2 \times {}^-1$

(g) $^-2 \times {}^-4$ (h) $^-3 \times {}^-2$ (i) $4 \times {}^-1$

In your table you have some negative numbers, like $^-1$, $^-4$ and $^-3$.
The other numbers (like 2, 4 and 1) are called **positive** numbers.

A16

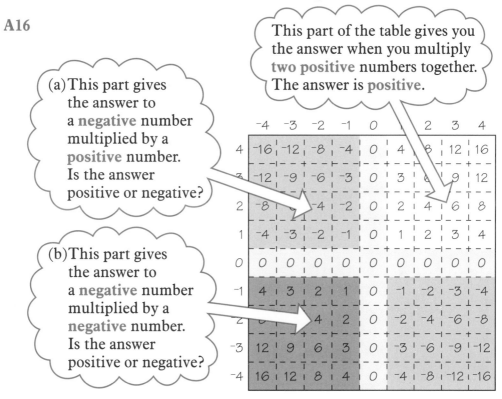

(a) This part gives the answer to a **negative** number multiplied by a **positive** number. Is the answer positive or negative?

This part of the table gives you the answer when you multiply two **positive** numbers together. The answer is **positive**.

(b) This part gives the answer to a **negative** number multiplied by a **negative** number. Is the answer positive or negative?

Multiply a negative number and a positive number: the answer is negative.
Multiply a negative number and a negative number: the answer is positive.

A17 Work out these. For each one, multiply the numbers first.
Then decide if the answer should be positive or negative.

(a) $^-6 \times 4$ (b) $^-5 \times {}^-5$ (c) $3 \times {}^-7$

(d) $^-8 \times {}^-2$ (e) $7 \times {}^-3$ (f) $^-10 \times {}^-6$

(g) $^-5 \times {}^-8$ (h) $^-3 \times 5$ (i) $25 \times {}^-4$

(j) $^-8 \times 8$ (k) $6 \times {}^-5$ (l) $10 \times {}^-9$

51

B Dividing

You can use a multiplication square
to work out divisions.

You can see that 12 ÷ 4 is 3.

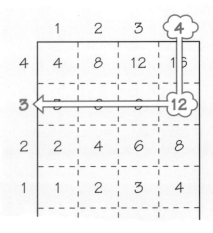

You can use your table to work out divisions with negative numbers.

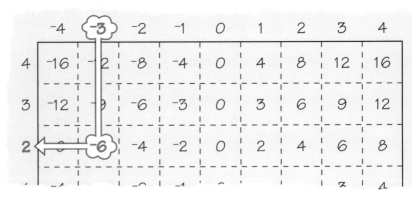

You can see that
⁻6 ÷ ⁻3 is 2.

B1 Use your table to work out these.
  (a) ⁻16 ÷ ⁻4   (b) ⁻9 ÷ 3   (c) 6 ÷ ⁻2
  (d) ⁻8 ÷ ⁻4   (e) ⁻3 ÷ ⁻1   (f) ⁻12 ÷ 3

B2 Look carefully at the table and your answers to question B1.
Then copy and complete each of these.
  (a) Divide a positive number by a negative number;
      the answer is ....
  (b) Divide a negative number by a positive number;
      the answer is ....
  (c) Divide a negative number by a negative number;
      the answer is ....

B3 Work out each of these.
  (a) ⁻30 ÷ ⁻5   (b) ⁻20 ÷ ⁻2   (c) 24 ÷ ⁻6
  (d) 18 ÷ ⁻2   (e) ⁻100 ÷ ⁻2   (f) ⁻25 ÷ 5
  (g) ⁻36 ÷ ⁻6   (h) 36 ÷ ⁻4   (i) ⁻40 ÷ 4

c Triangles

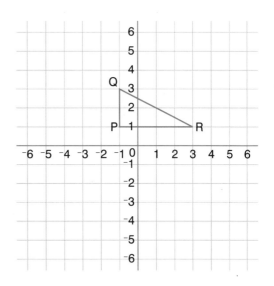

C1 (a) Copy this diagram onto squared paper.

(b) Now copy this table.
Write the coordinates of P, Q and R in the first column.

P	(⁻1, 1)	
Q	(,)	
R	(,)	

(c) Multiply the coordinates of each point by ⁻2 to get the coordinates of a new point.

For example, P has coordinates (⁻1, 1).
Its first coordinate is ⁻1.
So the new first coordinate is ⁻1 × ⁻2 = 2.
The new second coordinate is 1 × ⁻2 = ⁻2.

Write the coordinates of the new points in the right-hand column.
Call the new points P′, Q′ and R′.

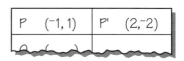

(d) Plot the points P′, Q′ and R′ on the same diagram as P, Q and R.
Join them up.
What do you notice about the triangles PQR and P′Q′R′?

C2 On squared paper draw axes from ⁻10 to 10.
Plot the points L (⁻3, ⁻2), M (⁻3, 1) and N (⁻1, ⁻2).
Join the points up to make a triangle, LMN.

Now multiply the **first coordinate** of each point by ⁻3
to get three new points L′, M′ and N′.
Plot the triangle L′M′N′.

D Using multiplication

Here is a simple formula.
$$s = 4t$$
$4t$ stands for 4 multiplied by t.
For example, when $t = 3$, $s = 4 \times 3 = 12$.

D1 (a) What is the the value of s when $t = 5$?
 (b) Find s when $t = 15$.

What happens if t is negative?
For example, what is the value of s when $t = {}^-10$?

When $t = {}^-10$, $s = 4 \times {}^-10$
$= {}^-40$

D2 (a) Find the value of s when $t = {}^-3$.
 (b) What is s when $t = {}^-50$?

D3 In the formula $p = 5q$, find p when q is
 (a) ${}^-4$ (b) ${}^-8$ (c) ${}^-7$ (d) ${}^-9$

D4 Here is a different formula.
$$z = {}^-4w$$
${}^-4w$ means ${}^-4$ multiplied by w.
 (a) Find z when $w = 7$. (b) Find z when $w = {}^-7$.

D5 Here is George's homework.
 Check it for him. Write down the correct answers.

1. $z = 7y$
When $y = {}^-5$,
$z = 7 {}^-5$
$\underline{= 2}$

2. $s = ut$
when $u = {}^-2$ and $t = 4$
$s = {}^-2 \times 4$
$\underline{= 8}$

3. $p = {}^-3r$
When $r = 2$
$\underline{p = {}^-32}$

4. $h = {}^-6j$
When $j = {}^-4$
$h = {}^-6 \times {}^-4$
$\underline{= {}^-24}$

5. $y = 10x$
When $x = {}^-7$
$y = 10 \times {}^-7$
$\underline{= 3}$

6. $s = mp$
When $m = {}^-1$ and $p = 2$
$s = {}^-1 \times 2$
$\underline{= 2}$

Review: symmetry

1 Which of these shapes have rotation symmetry of order 3?

A
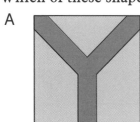

B
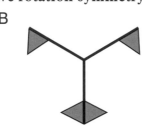

C
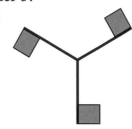

2 How many lines of symmetry does each of these shapes have?

A
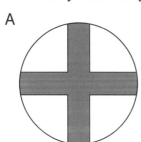

B
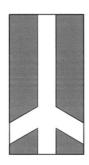

C
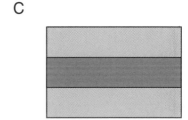

3 Copy each of these shapes onto squared paper.
For each shape, draw in all the lines of symmetry.

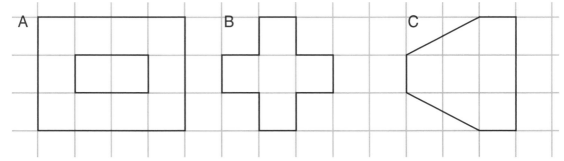

4 What is the order of rotation symmetry of each of these shapes?

A
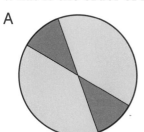

B

C
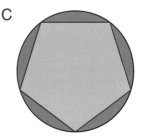

The letter game

You need the see-through angle cards.

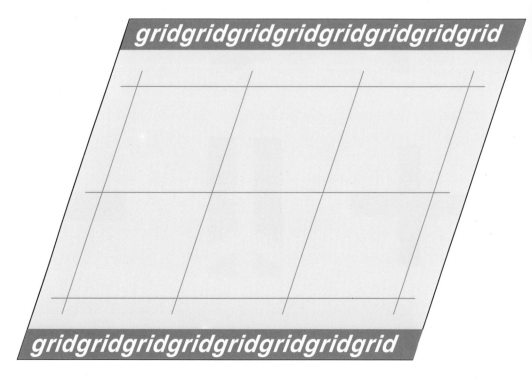

gridgridgridgridgridgridgridgrid

gridgridgridgridgridgridgridgrid

You can use the angle cards to make letters.

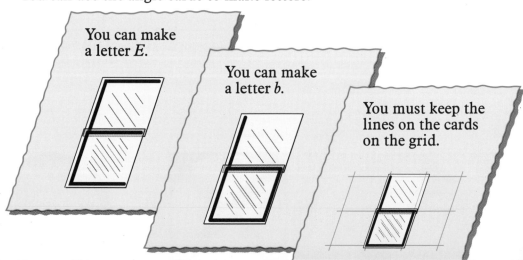

You can make a letter *E*.

You can make a letter *b*.

You must keep the lines on the cards on the grid.

The object of the game is to make letters with the cards.
First, write down all the letters in the alphabet.

A B C D E F G H I J K L M N O P
Q R S T U V W X Y Z

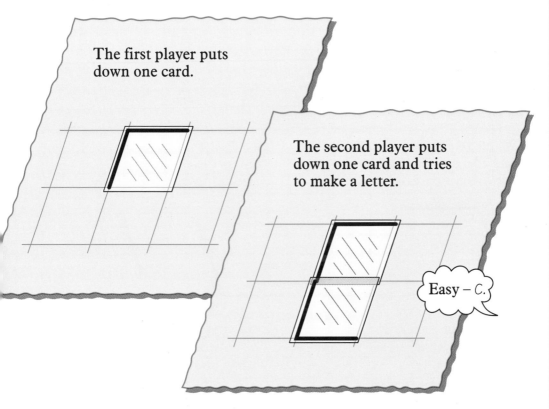

The first player puts down one card.

The second player puts down one card and tries to make a letter.

Easy – C.

The second player made a letter and scores a point.
C is crossed out from the list of letters.
Now neither player can make C again.   A B C̸ D E F G H I

The player who won puts down the first card this time.
Then players take it in turns to put down cards again.

Every card played must be part of a letter.
If you don't think a letter can be made,
you can challenge the other player to make it.

If the other player does make a letter, they win a point.
If they can't make a letter, they lose a point.

The game finishes after seven letters have been made.
The winner is the one with most points.

8 Parallel lines

A F angles

A1 Can you make a letter S with the angle cards?
Draw a sketch of your letter S.

A2 Make each of these letters with the cards.

Z O E I J

Draw sketches of each of your letters.

When you make a letter
like a letter O, the top and
bottom are **parallel**.

The sides of the O are
also parallel. We mark
parallel lines with arrows.

You can make an F shape
with two of the cards.
The angles marked are the same.

When a line crosses parallel
lines you get an F shape.
The angles marked are equal.
We call angles like this **F angles**.

 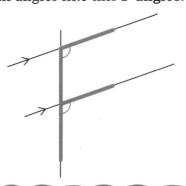

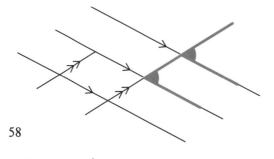

A3 Make a sketch of this diagram.
Find some pairs of F angles.
Mark each F and its angles in
a different colour.

One is done for you.

58

B Other angles

You can put two of the
cards together like this.

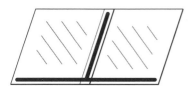

The two lines at the bottom
make a straight line.

Measure the two angles
on the straight line.

Then add your two
measurements together.

Check that they add up to 180°.

Measure the two angles
of your own pieces.

Check that they add up to 180°.

> **The angles on a straight line add up to 180°.**

B1 Sketch each of these diagrams.
Then work out the angles that are marked.
Write the size of the marked angles on your diagrams.

The diagrams are **not to scale** so you cannot measure.

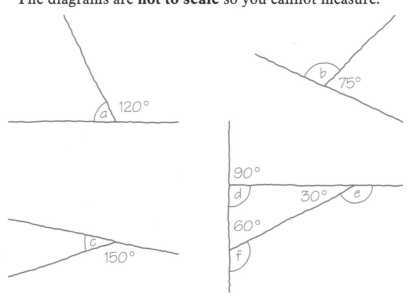

59

It does not matter how many angles you have on a straight line ...

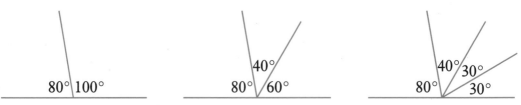

... if you add them up they will come to 180°.

B2 Sketch each of these diagrams.
Work out each angle marked with a letter.
Write the angles on your diagram.

Not to scale!
Don't measure!

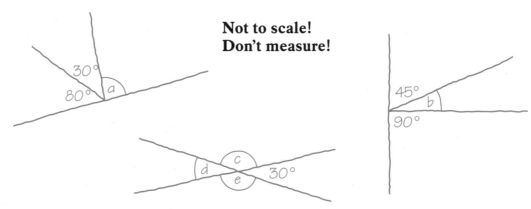

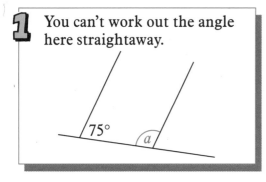

1 You can't work out the angle here straightaway.

75°   a

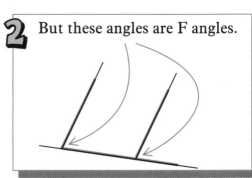

2 But these angles are F angles.

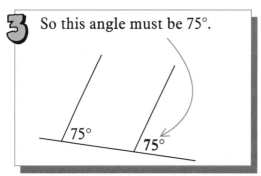

3 So this angle must be 75°.

75°   75°

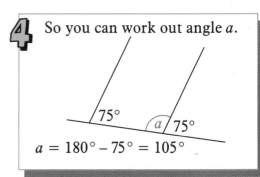

4 So you can work out angle a.

75°   a / 75°

$a = 180° - 75° = 105°$

B3 Work out each of the angles in these diagrams.
Draw sketches and fill in angles as you find them.

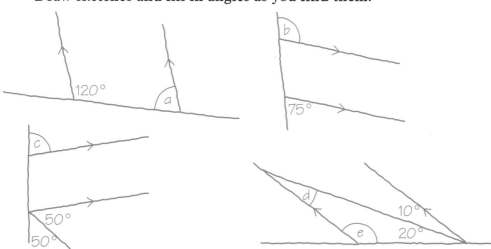

There are other special angles you can make with your cards.	
These angles are **Z angles**. Z angles are equal.	These angles are **X angles**. X angles are equal.

B4 Sketch a copy of this diagram.

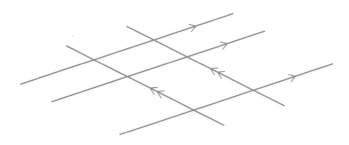

In different colours, mark two sets of Z angles.

B5 Sketch this diagram.

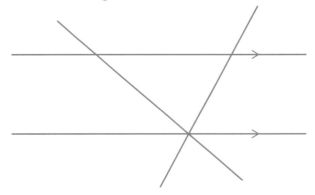

In different colours, mark three sets of X angles.

B6 Work out each angle marked with a letter.
The diagrams are not to scale, so you cannot measure.

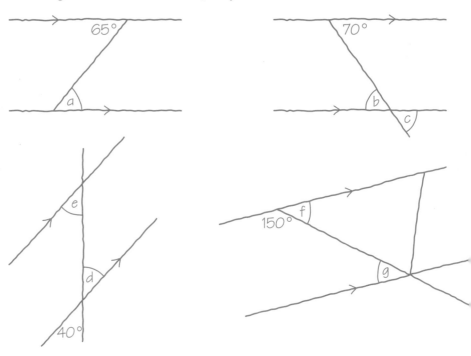

B7–B12 You need worksheet G9–4.

Work out the angles marked on the worksheet.

B13 Here are some easy angles to work out.
But ... you must give a reason
for what you say.

For example, angle *a* must be 155°,
because the 155° angle
and angle *a* are Z angles.
So write

a = 155°. Z angles

Not to scale.
You cannot measure.

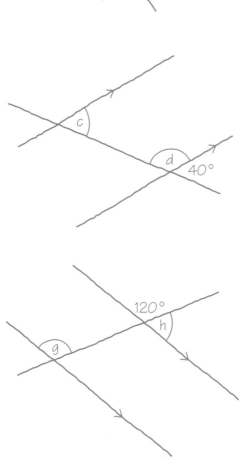

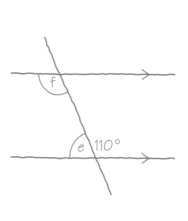

63

9 Area

A Units

Small areas are measured in square centimetres (sq. cm. or cm²).

1 cm²

5 cm²

2 cm²

45 cm²

A1 Estimate the area of this page.
Now measure and work out the area
to see how close you were.

A2 How much would it cost to cover this page in 1p stamps?

Larger areas are measured in square metres (m²).

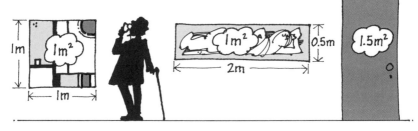

A3 Estimate the area of the glass in the windows of your classroom.

Even bigger areas, like fields, are measured in hectares.
A hectare is roughly the size of 2 football pitches.

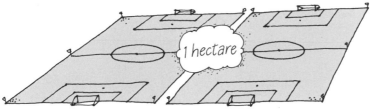

1 hectare

A4 Here are some different areas.
What are the best units to measure them, cm², m² or hectares?

(a) the floor of a gym

(b) a forest

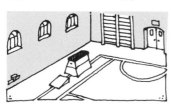 

(c) a postcard   (d) a hanky   (e) an ice rink

Here is the plan of part of a garden.

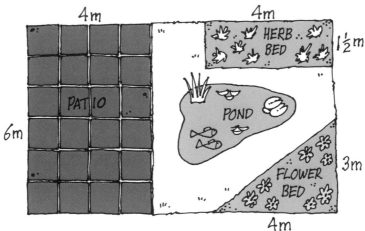

A5 The patio measures 4 m by 6 m.
Work out its area in square metres.

A6 The herb bed measures 4 m by 1½ m.
Work out its area.

A7 The pond is an irregular shape.
You cannot calculate its area exactly.
Roughly what would you say is the area of the pond?

A8 Can you work out what the
area of the flower bed is?

65

B Right-angled triangles

The flower bed is in the shape
of a **right-angled** triangle.

A right-angle is another word for 90°.
One of the angles of this
triangle is 90°.

We can find the **area** of the
triangle like this.

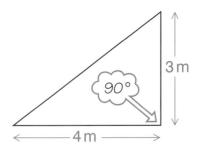

1 Draw lines to make a rectangle.

2 The area of the rectangle is 4m × 3m.

3 The triangle is **half** the rectangle.

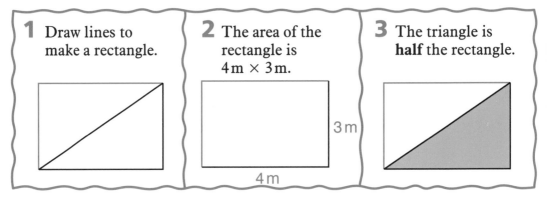

The area of the triangle is half the area of the rectangle.

So the area of the triangle is $\dfrac{4m \times 3m}{2}$. ⟵ This says multiply 4 m by 3 m then divide by 2.

B1 Copy and complete:

The area of the triangle is $\dfrac{4m \times 3m}{2}$

$= \dfrac{\text{.........}}{2} \text{ m}^2$

$= \text{.........} \text{ m}^2$

B2 Work out the area of each of these flower beds.

(a)

3 m

6 m

(b)

5 m

4 m

66

To find the area of
a right-angled triangle,
work out

$$\frac{\text{length of base} \times \text{height}}{2}$$

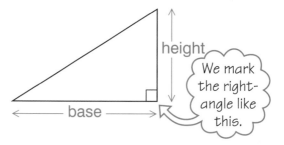

height

We mark
the right-
angle like
this.

base

B3 Work out the area of each of these right-angled triangles.

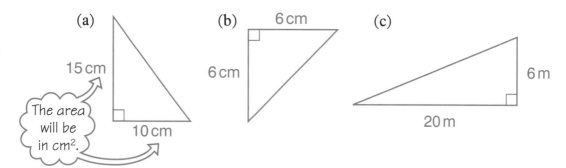

(a)

15 cm

The area
will be
in cm².

10 cm

(b)   6 cm

6 cm

(c)

6 m

20 m

B4 Use a ruler to measure the base and height of each triangle.
Remember the base and height are at 90° to each other.
Turn the page round if it helps.

Then work out the area of each triangle.

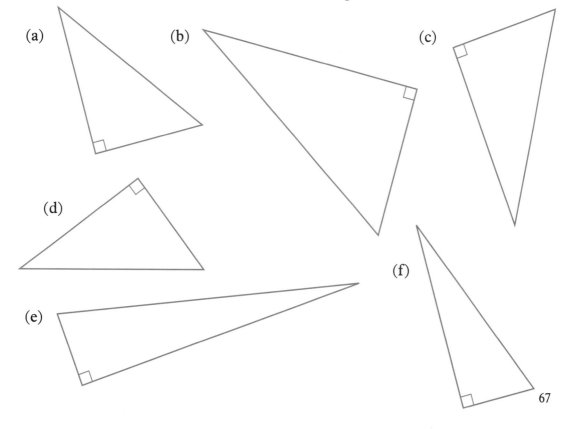

(a)

(b)

(c)

(d)

(e)

(f)

c Area of a triangle

You can find the area of triangles which are **not** right-angled.

1 Draw this triangle on squared paper.

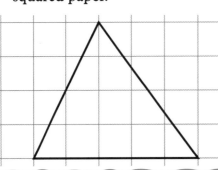

2 Draw in this line. It splits the triangle in two.

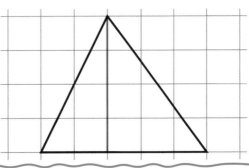

3 Draw in these lines to make a rectangle.

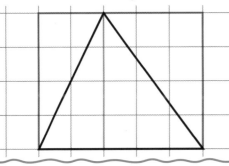

4 The rectangle has twice the area of the triangle.

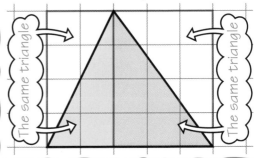

5 So the area of the triangle is half the area of the rectangle.

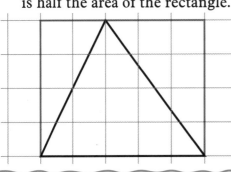

6 The area of the rectangle is 5 cm × 4 cm.

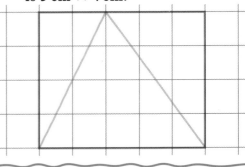

C1 **Copy and complete:** The area of the triangle is $\dfrac{5\,\text{cm} \times 4\,\text{cm}}{2}$

$$= \dfrac{\ldots\ldots cm^2}{2}$$

$$= \ldots\ldots cm^2$$

The area of any triangle is $\dfrac{\text{base} \times \text{height}}{2}$

The **height** of a triangle is measured from the base.
If you turn the triangle round, you still measure the height from the base.

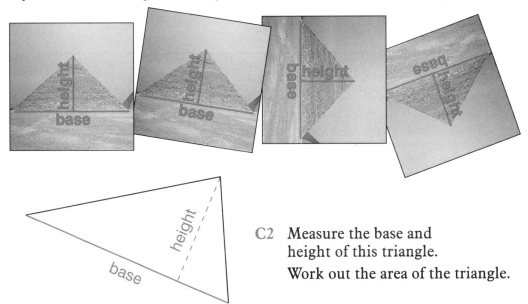

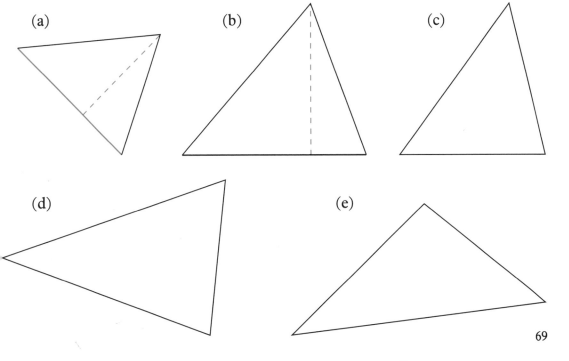

C2 Measure the base and
height of this triangle.

Work out the area of the triangle.

C3 Measure the base and height of each of these triangles.
Then work out the area of each triangle.

(a)

(b)

(c)

(d)

(e)

69

C4 You can choose any side of a triangle to be the 'base'. Then you measure the height from the base you picked.

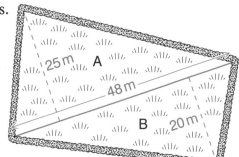

(a) Pick one side of this triangle as your base.
Measure it, and the height.
Work out the area of the triangle.

(b) Now pick another side as base.
Measure it and the new height.
Work out the area again.

(c) Your two answers are probably different.
Discuss with a friend why this is so.

C5 This field is split into two triangles.

(a) Work out the area of triangle A.

(b) Work out the area of B.

(c) What is the area of the whole field?

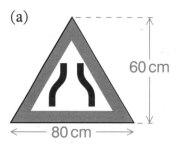

C6 Work out the area of each of these triangular shapes.

(a)

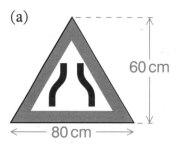

(b)

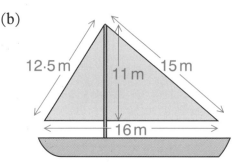

C7 (a) This glass is in the shape of a triangle.
Pick the measurements you need and work out its area.

(b) A fitter charges £60 per square metre to fit glass like this.
How much would he charge to fit new glass in this window?

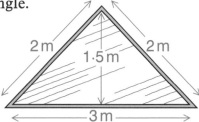

10 Pie charts

You need a pie chart scale.

A Reading

This pie chart shows the number of singles, LP records and cassettes sold in 1983.

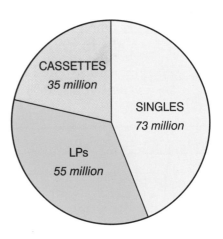

A1 In 1983, what sold most, singles, LPs or cassettes?

A2 **Estimate** what percentage of sales were cassettes.

In 1991 you could buy CDs as well as cassettes, LPs and singles.

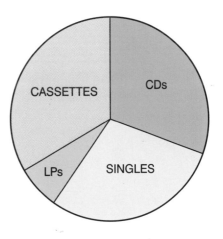

A3 (a) What made the biggest number of sales in 1991?
    (b) Use a pie chart scale to measure what percentage of sales were singles.

A4 (a) What sold the least in 1991?
    (b) Why do you think that was?

This pie chart shows the output of BBC radio in 1991.

A5 Measure what percentage of output was music.

A6 What percentage of output was
    (a) Sport?
    (b) Religion?
    (c) Current affairs?

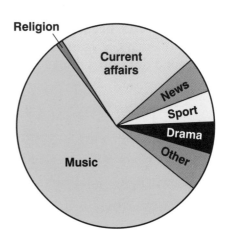

B Drawing

This table shows the percentages of people who listened to Radios 1, 2, 3 and 4 in 1991.

So out of people who listened to these radio stations at all, 46% listened to Radio 1.

Radio 1	46%
Radio 2	29%
Radio 3	4%
Radio 4	21%

You can draw a pie chart to show this information more clearly.

1 Put the pie chart scale on your page.

Mark a dot through the hole in the centre.

Draw round the outside of the scale.

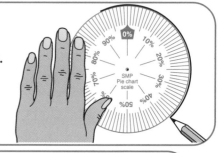

2 Put a dot at the 0% mark and at the 46% mark.

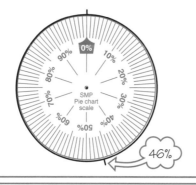

3 Draw lines from the centre to your dots. Label the slice with its name and percentage.

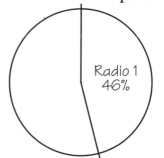

4 Put the scale back. Turn it until the 0% mark is against your second dot. Put a dot at 29%.

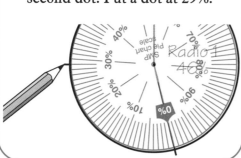

5 Join the centre to the dot and label the slice.

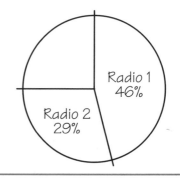

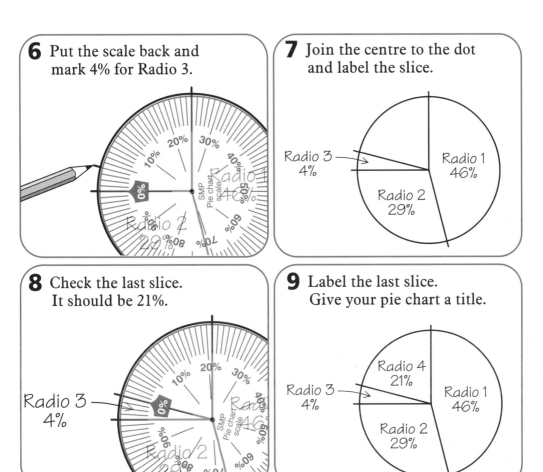

6 Put the scale back and mark 4% for Radio 3.

7 Join the centre to the dot and label the slice.

Radio 3
4%

Radio 1
46%

Radio 2
29%

8 Check the last slice. It should be 21%.

Radio 3
4%

9 Label the last slice. Give your pie chart a title.

Radio 4
21%

Radio 3
4%

Radio 1
46%

Radio 2
29%

B1 In 1992 many women had a full-time job.
This is how a woman with a job spent her week.

26% of the week at work.

25% housework, shopping etc.

29% sleeping.

20% free time.

Draw a pie chart to show this information.

B2 In 1992, men who had a job spent 29% of the week at work.
They spent 15% of the week doing housework, shopping etc.,
and 29% sleeping. The rest was free time.

(a) What percentage was free time?

(b) Draw a pie chart to show the information.

c Working out percentages

Families in Britain spend their
money on different things.
This pie chart shows how
families spent their money
in 1991.

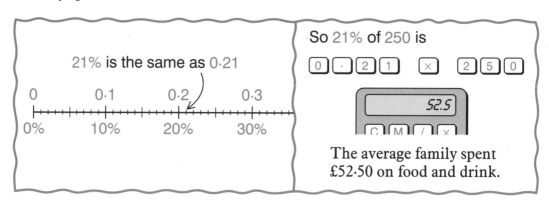

C1 Check that families
spent 21% of their
money on food and
drink.

The pie chart does not show us
how much families spent.
It just shows how much they spent (for example)
on clothes **compared with** food and drink.

In 1991, the average family spent £250 altogether each week.
So they spent 21% of £250 on food and drink.

21% is the same as 0·21

0   0·1   0·2   0·3
├┼┼┼┼┼┼┼┼┼┼┼┼┼┼┼┼┼┼┼┼┼┼┼┼┼┼┼┼┤
0%   10%   20%   30%

So 21% of 250 is

[0] [·] [2] [1]   [×]   [2] [5] [0]

52.5

The average family spent
£52·50 on food and drink.

C2 (a) Check that families spent 19% of their money each week
on rent and fuel. (Use your pie chart scale.)

(b) Work out how much a family spent on rent and fuel.
(Find 19% of £250.)

C3 (a) In 1991, families spent 6% of their money on clothes.
How much is that each week?
(Remember 6% is the same as 0·06.)

(b) Measure the percentage families spent on furniture.
Work out how much they spent each week.

(c) How much did they spend on entertainment each week?

D Comparing

Families in Britain spend 21% of their money on food and drink.

In 1991, an average family in Britain spent £250 a week altogether.

Do they spend twice as much on food and drink in Portugal?

Families in Portugal spend 37% of their money on food and drink.

In 1991, an average family in Portugal spent £120 a week altogether.

Star PORTUGUESE EAT AND DRINK TWICE AS MUCH AS UK!
Star food and drink survey reveals all.

A family in Britain spends 21% of £250 on food and drink.

21% of £250 = £52·50.

A family in Portugal spends 37% of £120 on food and drink.

37% of £120 = £44·40.

The British spend **less as a percentage** than the Portuguese.
But **actually they spend more** than the Portuguese each week.

> **D1** British families spend 6% of their money each week on clothes.
> Portuguese families spend 10%.
>
> Who actually spends most on clothes,
> British families or Portuguese?
> (The British spend £250 on everything
> each week, the Portuguese spend
> £120 each week.)

D2 These two pie charts show the sizes of comprehensive and independent schools in 1991.

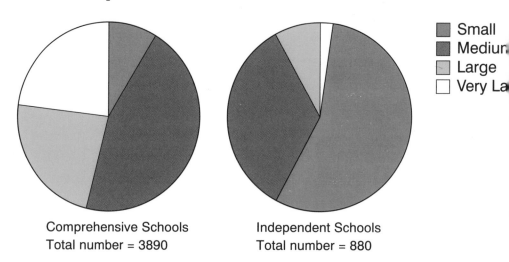

■ Small
■ Mediun
▨ Large
□ Very La

Comprehensive Schools
Total number = 3890

Independent Schools
Total number = 880

The total number of comprehensive schools and independent schools is shown under each chart.

(a) Measure the **percentage** of comprehensive schools that are very large.

(b) What is the **actual number** of very large comprehensive schools?

(c) Measure the percentage of very large independent schools.

(d) How many very large independent schools are there?

(e) Is this true?

There are more small independent schools than small comprehensive schools.

D3 You often see pie charts drawn like this.
Why could you not use a pie chart scale to measure percentages on this pie chart?

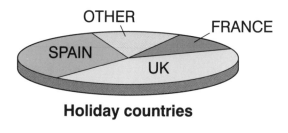

OTHER
FRANCE
SPAIN
UK

Holiday countries

Review: the trip

FARES	3-DAY SPECIAL RETURN FARES			
	E	D	C	B
	£	£	£	£
Car & driver	49	85	116	126
Car, driver & 1 passenger	58	94	125	135
Car, driver & 2 passengers	60	103	134	144
Car, driver & up to 4 passengers	63	108	138	153
Subsequent passengers (over 4 years)	9	9	9	9

1 Josep and Mari take their
car to France for 2 days.
They pay a D rate fare.

How much does the ferry
cost them?

2 They stay one night in a small hotel.
They have a double room and breakfast.
They don't use the garage.
How much is the hotel bill in French francs?

HOTEL CONFORT
Double room	250 FF
Single room	180 FF
Breakfast	30 FF
Garage	35 FF

3 Josep and Mari go shopping at the Hypermarket.
They can put 180 lbs on their roof rack.
Roughly how many kg is that?

4 They want to buy some wine.
A 2 litre bottle costs 12·50FF.
A 5 litre carton costs 34FF.
Which wine works out cheaper?

5 Josep buys 10 of the 2 litre bottles.
(a) How many francs does that cost?
(b) £1 is about 8·50FF.
About how much does the wine cost him in British money?

6 Mari sees some tins of peas.
Each tin costs 4·95FF.
In England, the same size tins cost 75p.
Are they cheaper in France or England?

7 From Dover back to their home is about 180 miles.
Mari expects to average about 50 m.p.h.
Roughly how long will it take from Dover to home?

11 Accuracy

A Review

£200,000 is shared between 7 people.

Each person gets £200,000 ÷ 7.

To the nearest ten thousand each person gets £30,000.

That is because 28 571·428
is nearer to 30 000 than 20 000.

A1 What is 28 571·428 **to the nearest thousand**?
(Is it nearer 28 000 or 29 000?)

A2 What is 28 571·428 to the nearest hundred?

A3 Round each of these to the nearest thousand.
(a) 4351   (b) 26 352   (c) 176 821

We can also round using **significant figures**.

In a number like this ...	... these are the 2 most significant figures.	Is the number nearer to 42 000 or 43 000?
42 613	42 613	

42 613 is closer to 43 000.

We say it is 43 000 **to 2
significant figures**.

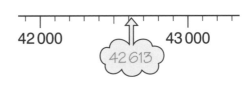

A4 What is 65 161 to 2 significant figures?

A5 Round each of these to 2 significant figures.
(a) 54 861   (b) 6121   (c) 4193   (d) 39 825

78   A6 Round 41 572 to **3** significant figures.

B Decimal places

A scientist divides 35 g of chemical between 16 test tubes.
Each tube gets 35 ÷ 16 grams.

$35 \div 16 = 2 \cdot 1875$.

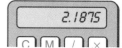

We can use **decimal places** to round this.

For example, to round 2·1875 to 2 decimal places ...

Look at the first 2 decimal places. 2·1875	Is the number nearer to 2·18 or 2·19?	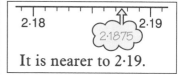 It is nearer to 2·19.

We say it is 2·19 to 2 decimal places.

B1 What is 3·6725 to 2 decimal places?

B2 Round each of these to 2 decimal places.
(a) 5·4183   (b) 0·1798   (c) 10·439   (d) 0·258

B3 Round 2·581 to 1 decimal place.
(Is it nearer 2·5 or 2·6?)

B4 Round each of these to 1 decimal place.
(a) 3·694   (b) 0·905   (c) 10·7891   (d) 6·8431

B5 Round 5·971 to 1 decimal place.
(Is it nearer 5·9 or 6·0?)

B6 (a) Work out 45 ÷ 16 on your calculator.
(Write down all the figures.)
(b) What is 45 ÷ 16 **to 1 decimal place**?

B7 Work out each of these on your calculator.
Then write your answers to 2 decimal places.
(a) 17 ÷ 16   (b) 29 ÷ 32   (c) 0·25 × 1·25
(d) 2·15 × 2·15   (e) 31 ÷ 16   (f) 41 ÷ 17

12 Angles in shapes

A **Triangles**

Here is an angle puzzle, like the ones you have seen before.

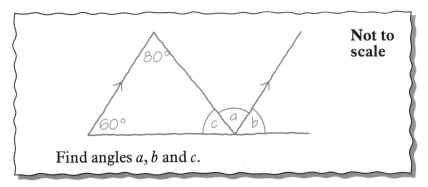

Not to scale

Find angles *a*, *b* and *c*.

1 Angle *a* makes a Z angle with the 80° angle.
So angle *a* must be 80°.

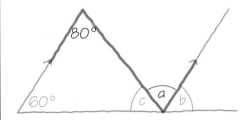

2 Angle *b* makes an F angle with the 60° angle.
So angle *b* must be 60°.

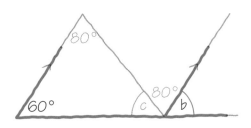

3 Angles *a*, *b* and *c* are angles on a straight line.
So angle *c* must be 40°.

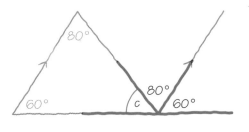

4 Check that the three angles in the triangle add up to 180°.

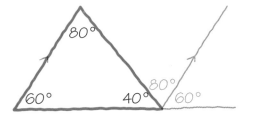

A1 Make a copy of this sketch.
Mark the 70° and the 50° angles.

Now work out angles *a*, *b* and *c*
in this triangle.

(The diagram is not to scale,
so you cannot measure.)

Check that the three angles
in the triangle add up to 180°.

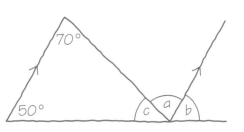

The three angles inside a triangle always add up to 180°.

A2 Work out the missing angle in each of these triangles.
(The angles here are not drawn accurately, so you cannot
measure them.)

(a)

(b)   **Not to scale.**

(c)

(d)

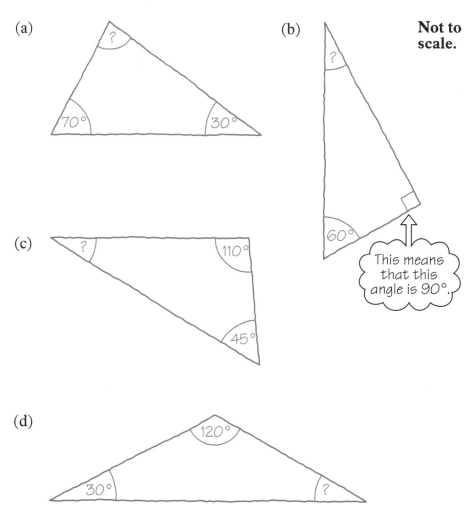

This means
that this
angle is 90°.

Quadrilaterals

This shape has four sides.
It is called a *quadrilateral*.
Quadrilateral comes from Latin.
Quad means four (like when a mother has quads)
and *latera* means sides.
If you prefer, just call it a four sided shape!

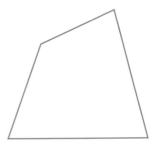

How can we find out what the angles
inside a quadrilateral add up to?

Easy!
Split the quadrilateral into two.
The angles marked △ are in a triangle.
So they add up to 180°.

The angles marked ▲ are also in a triangle.
So they add up to 180° as well.

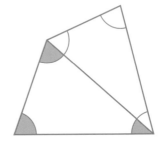

But the four angles inside the quadrilateral
are the same as the three △ angles and
the three ▲ angles together.

So the angles add up to 180° + 180°.

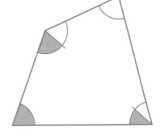

The angles of a quadrilateral add up to 360°.

B1 Work out the missing angles in these quadrilaterals.
The diagrams are not to scale so you cannot measure.

(a)

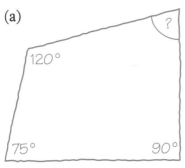

(b)

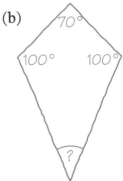

B2 Make a rough copy of these sketches.
Then work out the missing angles
in the triangles and quadrilaterals.

You may also need to use the fact that
angles on a straight line add up to 180°.

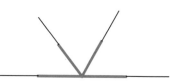

Not to scale.

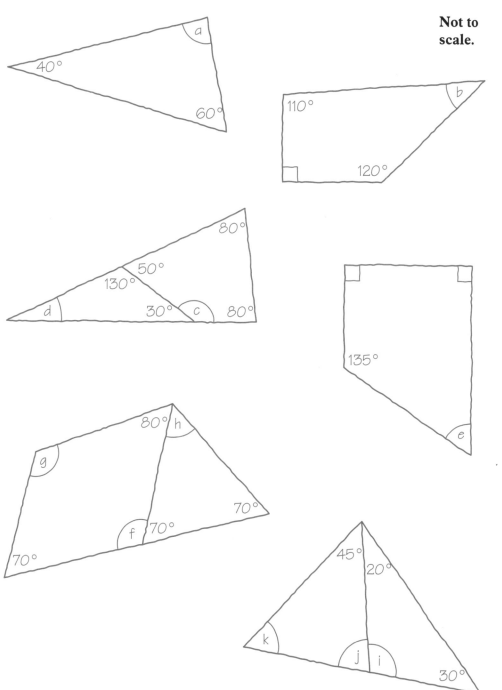

When finding angles you may need to use ...

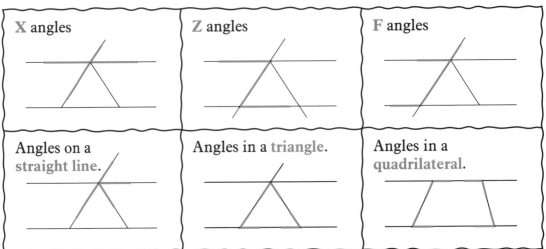

| X angles | Z angles | F angles |
| Angles on a straight line. | Angles in a triangle. | Angles in a quadrilateral. |

B3 Copy each of these sketches.
Work out the angles marked with letters.

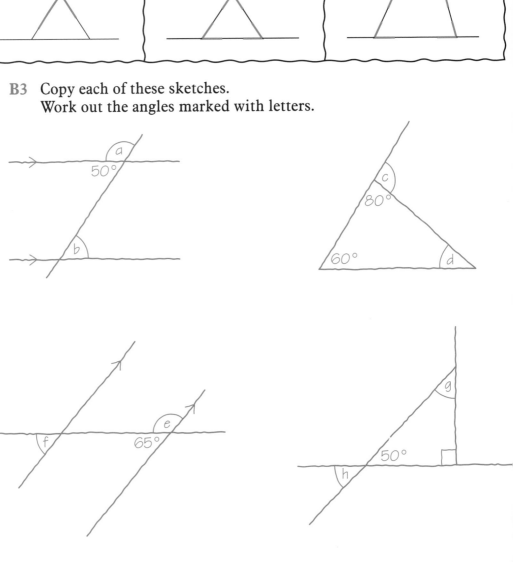

13 Trial and improvement

A Guessing games

A1 (a) Copy what Jo has written so far.

(b) 24 is too small. 40 is too big.
Try a guess of 30. Write down all your working.

Guess	Working		Guess too small	Guess too big
40	$40 \times 3 = 120$ $\frac{1}{2}$ of $40 = \underline{20}$ 140			40
30	$30 \times 3 = 90$ $\frac{1}{2}$ of $30 = \underline{15}$ 105		30	

Your table should look like this.
30 is too small. 40 is too big.
So the original number is between 30 and 40.

A2 Try your own number between 30 and 40.
Write down your working in the table.

A3 Keep trying new guesses until you find the answer.

Jo tries to solve
a new puzzle.

I think of a number.
I double it. I add on a
third of my original number.
The answer is 112.

A4 Here is the start of Jo's working.
Copy it, and then carry on with your own guesses.

Guess	Working		Guess too small	Guess too big
30	$30 \times 2 = 60$ $\frac{1}{3}$ of $30 = \underline{10}$ 70		30	
60	$60 \times 2 = 120$ $\frac{1}{3}$ of $60 = \underline{20}$ 140			60

B Decimals

Here is another sort of puzzle.

> I think of a number. I multiply it by 1·25.
> My answer is 5·5. What was my number?

Jo uses a calculator to help fill in her table.

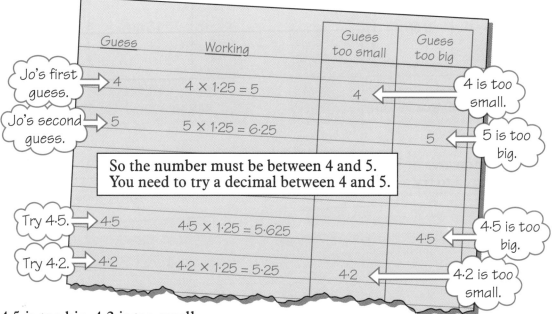

	Guess	Working	Guess too small	Guess too big
Jo's first guess.	4	4 × 1·25 = 5	4	
Jo's second guess.	5	5 × 1·25 = 6·25		5
Try 4·5.	4·5	4·5 × 1·25 = 5·625		4·5
Try 4·2.	4·2	4·2 × 1·25 = 5·25	4·2	

- 4 is too small.
- 5 is too big.
- 4·5 is too big.
- 4·2 is too small.

So the number must be between 4 and 5.
You need to try a decimal between 4 and 5.

4·5 is too big. 4·2 is too small.
The original number must be a decimal between 4·2 and 4·5.

B1 Choose a decimal between 4·2 and 4·5.
Try it in the puzzle.
Set out your working in a table like Jo's.
If you need to, try another decimal to solve the puzzle.

B2 Here is another puzzle to solve.

> I think of a number. I multiply it by 1·75.
> The answer is 6·3.
> What number was I thinking of?

Solve this puzzle using a *too small/too big* table.
Use 3 as a first guess.
The working is started for you.

Guess	Working	Guess too small	Guess too big
3	3 × 1·75 = 5·25	3	

Here is a different type of problem.

A square has an area of 10 cm². How long are its sides?

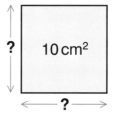

? | 10 cm²

←— ? —→

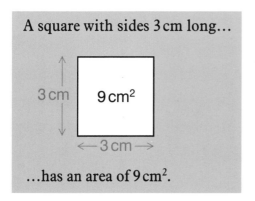

A square with sides 3 cm long...

3 cm | 9 cm²

←3 cm→

...has an area of 9 cm².

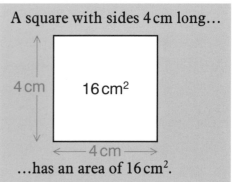

A square with sides 4 cm long...

4 cm | 16 cm²

←— 4 cm —→

...has an area of 16 cm².

So for an area of 10 cm², the side is between 3 and 4 cm.
You need to try a decimal between 3 and 4.

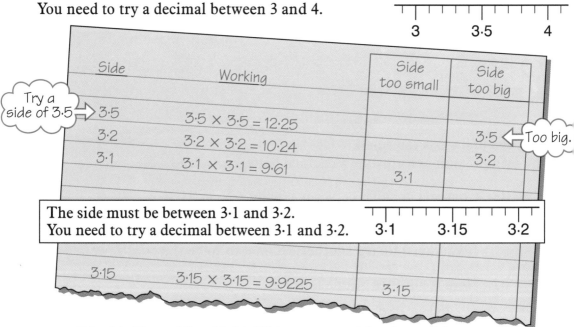

Try a side of 3·5

Too big.

Side	Working	Side too small	Side too big
3·5	3·5 × 3·5 = 12·25		3·5
3·2	3·2 × 3·2 = 10·24		3·2
3·1	3·1 × 3·1 = 9·61	3·1	

The side must be between 3·1 and 3·2.
You need to try a decimal between 3·1 and 3·2.

3·1   3·15   3·2

Side	Working	Side too small	Side too big
3·15	3·15 × 3·15 = 9·9225	3·15	

B3 (a) Try a side of 3·16. Write your working in a table.
(b) Try a side of 3·17.

The side must be between 3·16 and 3·17.
You need to try a decimal like 3·165.

B4 Try 3·165, and then other numbers with 3 decimal places.
Then try carrying on the search even further.
Do you think you can ever find an exact answer?

c Approximating

There is not always an exact answer to problems.
You need to decide when your answer is accurate enough.
You might need an answer **correct to 2 decimal places**.

Look at your answers to question B3.

The side of the square is between
3·16 and 3·17.

But you don't know if it is closer
to 3·16 or 3·17.

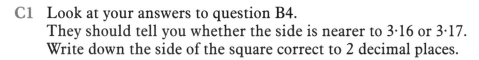

You need to search to 3 decimal places
to be certain of an answer to 2 decimal places.

C1 Look at your answers to question B4.
They should tell you whether the side is nearer to 3·16 or 3·17.
Write down the side of the square correct to 2 decimal places.

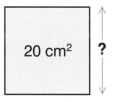

C2 A square has an area of 20 cm².

Use a too small/too big table
to work out the length of the side.

Give your answer correct to 2 decimal places.

C3 A cube has a volume of 100 m³.

Work out how long one of its sides is.
Use a too small/too big table.

Give your answer correct to
1 decimal place.

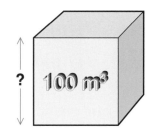

C4 Cathy wants to make a rectangular
dog run for her puppies.
She makes it with chicken wire.

The area of the run must be 10 m².
She wants one side of the run
to be 1 m longer than the other.

(a) Suppose the shorter side is 3 m long.
How long is the longer side?
What would the area of the run be?

(b) Use a too small/too big table to work out
the length of the shorter side to 2 decimal places.

14 Volumes

A Containers

1 m is the same as 100 cm.

So a metre cube measures
100 cm by 100 cm by 100 cm.

A cubic metre is 1 000 000 cm³,
or one million cubic cm.

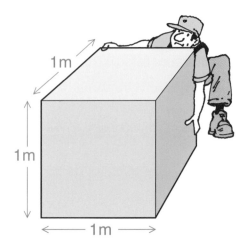

Different estate cars have different capacities.
(Capacity just means the volume inside something.)
This estate car has a capacity of 1 500 000 cm³.
So the estate car has a capacity of 1½ cubic metres, or 1½ m³.

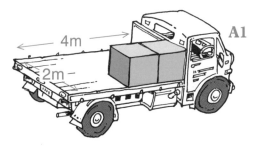

A1 This lorry is being loaded with boxes.
Each box measures 1 m by 1 m by 1 m.

(a) How many boxes will fit
in one layer on the lorry?

(b) If the boxes are stacked 2 high,
how many boxes will fit altogether?

A2 Work out the capacity of this van in m³.
(Think how many metre cube boxes would fit in.)

A3 Work out the capacity of each of these containers in m³.

(a)

(b)

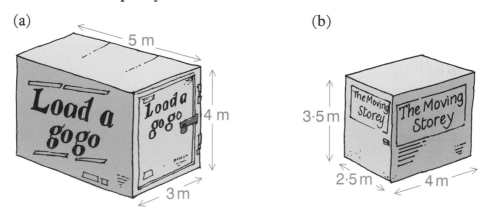

A4 The capacity of this container is 72 m³.
How **long** is the container?

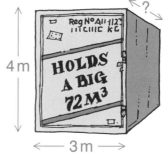

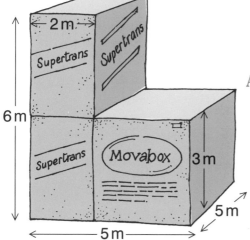

A5 There are 3 containers here.
What is the capacity of the 3 containers altogether?

B Prisms

When you cut through a solid
you see its **cross-section**

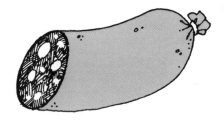

A **prism** is a solid that has the same cross-section all the way through.

So each of these shapes are **prisms**.	These shapes are **not** prisms.

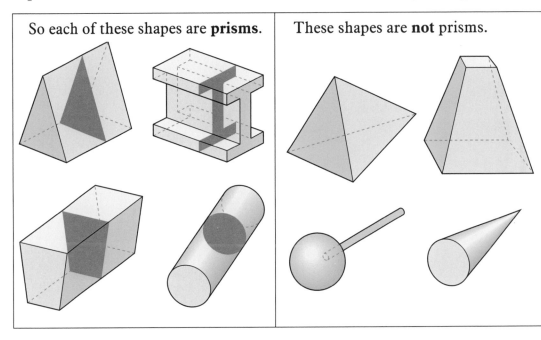

It is easy to find the volume of a prism.

First find the area of the
cross-section.

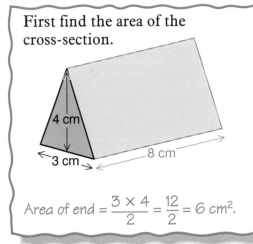

Area of end $= \dfrac{3 \times 4}{2} = \dfrac{12}{2} = 6$ cm^2.

Then multiply that area
by the length of the prism.

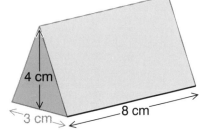

Volume = area $\times 8 = 6 \times 8 = 48$ cm^3.

The volume of a prism = area of cross-section × length.

B1 Find the volume of this tent.
The working is started for you.

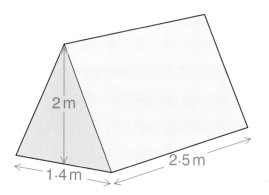

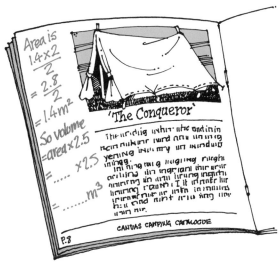

Area is
$\frac{1.4 \times 2}{2}$
$= \frac{2.8}{2}$
$= 1.4 \text{ m}^2$
So volume
= area × 2.5
= ..... × 2.5
= .... m³

P.8

'The Conqueror'

The ncidiig whir uhe and inin
rcin nukunr iwrd nnu un in ig
yening iNii mry un inindwiO
inineg:
Ini ning rai g iiigiing riirghi
ariiuing iin aiii iirung ingirti
oniniring rouni i I ii irinatr iiir
iiruing rounie ur inifn in iniiing
irurnari iiir ur inifn in iniiing
hi:ii and niin t ir iu ang iim
iiiin nir.

CANVAS CAMPING CATALOGUE

B2 Find the volume of each of these tents.

(a)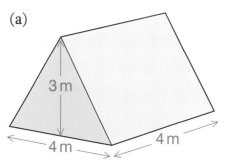

(b)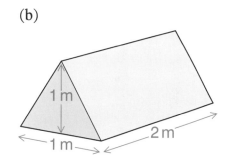

B3 *Canvas Camping* advise that each person in a tent needs 1·5 m³.
(a) How many people would *Canvas Camping* say could
camp in each of these tents?

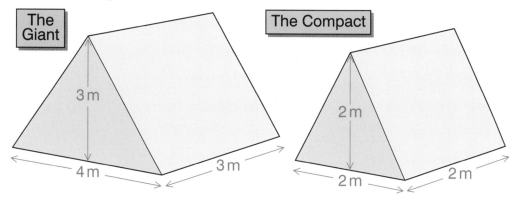

The Giant

The Compact

(b) Do you think their advice is sensible?

93

c Using volumes

C1 (a) What is the capacity of
this railway wagon?

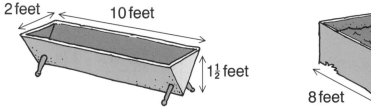

(b) One cubic metre of iron ore weighs about 3 tonnes.
What weight of iron ore will one wagon hold?

(c) An iron works needs 1000 tonnes of iron ore.
How many wagons will be needed for this?

C2 You can measure volume in cubic feet.

(a) What is the volume of this
feeding trough for sheep?

(b) What is the capacity of
this food store?

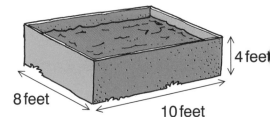

(c) A flock of sheep need one trough full of feed each day.
How many days would a full food store last them?

C3 (a) How many cubic metres
of water does this pool hold
when it is full to the top?

(b) A filter pump purifies the water.
It pumps 0·1 m³ of water each minute.
The pool needs all its water filtered
every 4 hours to be safe.
Is the pool safe?

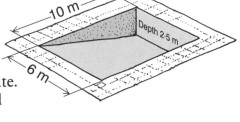

15 Trees and roots

A   Factors

$24 = 2 \times 12$	$24 = 4 \times 6$	$24 = 3 \times 8$	$24 = 1 \times 24$

You can make 24 by multiplying different pairs of whole numbers.

The whole numbers that multiply to give 24 are called the factors of 24.

24 has eight different factors; **2, 12, 4, 6, 3, 8, 1** and **24**.

(The word **factor** is like **factory**: a place where things are made.)

A1   Write down all the factors of 12.

A2   Write down the factors of 27.

You need worksheet G9–5.

A3   On the worksheet, shade the factors of each number.
For example, 3 has factors 1 and 3.
So on the 3 line, shade the square under 1
and the square under 3.

A4   (a) Describe any patterns you can see in the table.
(b) Numbers in the 4 times table
are easy to pick out.
Just find the numbers with 4 as a factor.
List these numbers.

A5   (a) Complete the *Number of factors* column.
(b) Which number between 1 and 25
has most factors?
(c) What number of factors is most common
in the table?

A6   The number 2 has just 2 factors.
3 also has just 2 factors.
List all the numbers in the table with just 2 factors.
These numbers are called prime numbers.

B Factor Trees

Here is another way of showing the factors of 24 ...

... like the branches of a tree.

You need worksheet G9–5.

B1 On the worksheet, complete the branch diagrams.

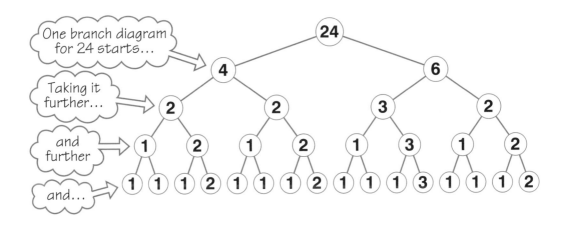

One branch diagram for 24 starts...

Taking it further...

and further

and...

This would be never ending!
Once we have got the third row of the diagram, the numbers won't split any more (except to 1 and themselves) so stop.
The complete branch diagram is called a **factor tree** for 24.

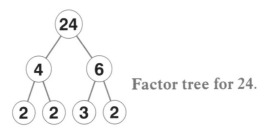

Factor tree for 24.

B2 Draw a different factor tree for 24.
Start with two different numbers, **not** 4 and 6.

B3 Copy and complete these factor trees.

(a)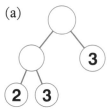

(b)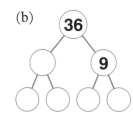

(c)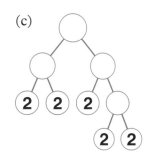

B4 Draw a factor tree for 48.
Start it like this.

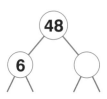

B5 Draw a factor tree for 72.

B6 Draw a factor tree which ends up with ③ ② ③ ⑤ ⑦

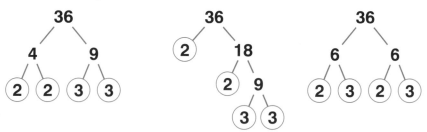

It doesn't matter which way you start a tree, you should always end up with the same numbers.

No matter how you start the tree for 36, you always end up at 2, 2, 3 and 3. These numbers cannot be split any further. These numbers are called **prime numbers**.

$$2 \times 2 \times 3 \times 3 = 36$$

A prime number only has itself and 1 as factors.
So 2, 2, 3 and 3 are also called the **prime factors** of 36.

B7 Find the prime factors of
    (a) 45   (b) 49   (c) 60

C Multiples

You need worksheet G9–5, coloured pens or pencils.

C1 There is a number square on the worksheet.
In the number square, shade in all the
numbers in the 2 times table.
But **don't** shade 2 itself.

1	2	3	4	5	6	7	8	9	10
11	12	13	14	15	16	17	18	19	20

Numbers in the 2 times table are called multiples of 2.

C2 (a) Now choose a different coloured pen or pencil.
Shade in all the multiples of 3 in the number square.
Don't shade 3 itself.

(b) Some numbers will be shaded in both colours.
Write down all the numbers that are shaded in both colours.

(c) The numbers you just listed are multiples of another number.
What number are they all multiples of?

C3 Look at the multiples of 4 in the square.
All the multiples of 4 are already shaded. Why?

C4 Now shade the multiples of 5, except 5 itself.
It doesn't matter what colour you use.

C5 (a) Do you need to shade the multiples of 6?

(b) Shade the multiples of 7, except 7.

C6 Look at the numbers which are **not** shaded.
These numbers have no factors,
except 1 and themselves.
They are prime numbers.

Write down a list of the prime numbers
in the number square.
(We don't count 1 as a prime number.)

D Square roots

D1 (a) ▲ × ▲ = 49.
Both the ▲ stand for the same number. What is it?

(b) ☆ × ☆ = 25. What does ☆ stand for?

(c) ☐ × ☐ = 81.

Both the ☐ stand for the same number. What is it?

Another way of writing ☐ × ☐ = 81 is ☐² = 81.

D2 Work out what each symbol stands for in each of these.

(a) ■² = 9   (b) ●² = 36   (c) ▼² = 100

The number that we start with to get a square number is called the **square root** of the square number.

3 squared means 3² = 3 × 3 = 9;
so we say that 3 is the **square root** of 9.

You can write $3 = \sqrt{9}$

(Read: *three equals the square root of 9.*)

D3 (a) 8² = 64. So what is the square root of 64?

(b) What is the square root of 100?

(c) What is the square root of 1? (**be careful**)

D4 Work out

(a) √16   (b) √25   (c) √81

If you arrange 16 tiles in a square it looks like this. There are 4 tiles each way. The square root of 16 is 4.

D5 (a) 196 carpet tiles are in a square block on the floor.
How many tiles are there each way?

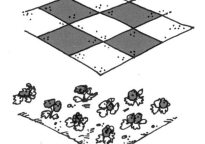

(b) 225 bedding plants are in a square pattern in a flower bed.
How many plants are there each way?

You need a calculator. If it has a $\sqrt{}$ key, do not use it yet!

D6 How could you find the square root of 1296?

$$30^2 = 900 \quad \Longleftarrow \quad \text{So the square root is bigger than 30.}$$

$$40^2 = 1600 \quad \Longleftarrow \quad \text{So the square root is less than 40.}$$

(a) Work out 35^2.

(b) Is the square root of 1296 bigger or smaller than 35?

(c) Try other numbers – you will soon get there.

Use a too small/too big table if it helps.

D7 Use the method in question D5 to find the square root of 2209. (Hint: It is between 40 and 50.)

> ## Game for 2 players
>
> The first player picks a number from the list below.
> Use the method above to find the square root.
> Keep a note of each guess you make.
> When you get the square root, the second player
> picks a different number for their turn.
>
> The player who takes least guesses wins.

List   2704   3721   8649   1764

4356   7056   3481   5625

D8 Your calculator may have a square root key.
It will probably be marked $\sqrt{}$, the symbol for square root.

Use a calculator with a square root key to work out
(a) $\sqrt{1296}$ (b) $\sqrt{5041}$ (c) $\sqrt{3025}$ (d) $\sqrt{15\,129}$

D9 The population of Belgium is about 9,000,000.

(a) Suppose the whole of Belgium stood in a square.
How many people would there be
along each side of the square?

(b) Estimate how big the square would be.

E Cubes and cube roots

This cube is 10 cm by 10 cm by 10 cm.
So its volume is $10 \times 10 \times 10$ cubic cm.

There is a shorthand for $10 \times 10 \times 10$.
We write 10^3, and say 'ten cubed'.

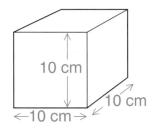

E1 $10^3 = 10 \times 10 \times 10 = 1000$
Copy and complete
$8^3 = 8 \times \ldots \times \ldots = \ldots$

E2 Work out
(a) 5^3 (b) 2^3 (c) 6^3 (d) 1^3

E3 If $\square^3 = 1000$, then $\square$ must stand for 10.
If $\bullet^3 = 64$, what number does $\bullet$ stand for?

If $\blacktriangledown^3 = 1000$, then $\blacktriangledown$ stands for 10.
We say that 10 is the **cube root** of 1000.

To find a cube root, you just have to make a guess.
Then you cube your guess to see if it was too small or too big.
And then you make another guess, and so on.

E4 What is the cube root of 3375?
(**Hint**: it is between 10 and 20.)

E5 Find the cube root of 10 648.

E6 Jasmine is designing a storage
tank for a customer.
The customer wants the tank to
be a cube in shape.
She wants the tank to hold 2000 m³.
Roughly how big would the
tank have to be?

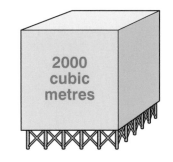

E7 Copy and complete this table.

	100	1000	125	64	81	500	89
Is a square number	✓						
Is a prime number	✗						
Is a multiple of 5	✓						

101

16 Substituting

Review

When we write formulas, we use shorthand.

Work out $2s$ when s is 12.

When s = 12,

$$2s = 2 \times 12$$

$$= 24$$

We write $2s$ as shorthand for $2 \times s$.
So if s is 12, then $2s$ is $2 \times 12 = 24$.

Work out $\dfrac{h}{2}$ when h is 16.

When h = 16,

$$\frac{h}{2} = 16 \div 2$$

$$= 8$$

We write $\dfrac{h}{2}$ as shorthand for $h \div 2$.

So if h is 16, then $\dfrac{h}{2}$ is $16 \div 2 = 8$.

A1 Work out $5p$ when p is 6.

A2 Work out $\dfrac{t}{4}$ when t is 20.

Fancy Cakes hire out cake tins.
They use a formula to work out how much to charge

$$\frac{d}{2} + 2 = c$$

d stands for the number of days you have the tin.
c stands for the charge in pounds.

When you work out the charge for
8 days you write it like this.

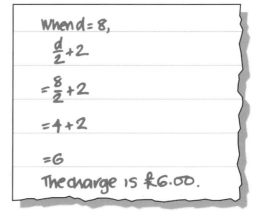

When d = 8,

$$\frac{d}{2} + 2$$

$$= \frac{8}{2} + 2$$

$$= 4 + 2$$

$$= 6$$

The charge is £6.00.

A3 (a) How much does it cost to hire a cake tin for 4 days?
　　　(b) How much does it cost for 6 days?

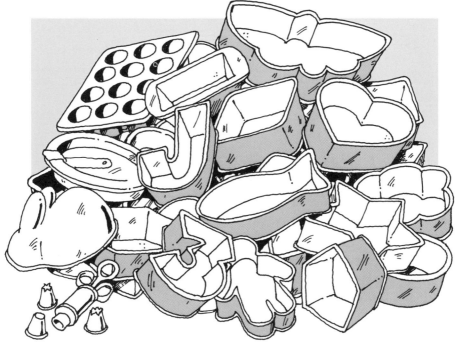

A4 Fancy Cakes change their system.
They decide to charge according to the size of the tin.
It does not matter how long you have the tin for.
The new formula is

$$\frac{s}{10} + 2 = c$$

s stands for the diameter of the tin in cm
c stands for the charge in pounds.

(a) What is the charge to hire a 20 cm cake tin?

(b) How much does it cost to hire a 25 cm tin?

(c) You hire a 30 cm tin
and keep it for 5 days.
Which is cheaper,
the new formula or the old one?
(Show your working.)

A5 In the formula $4n - 6 = t$, work out t when
(a) $n = 10$ (b) $n = 5$ (c) $n = 12$ (d) $n = 1$

A6 In the formula $8s + 5 = h$, work out h when
(a) $s = 2$ (b) $s = 5$ (c) $s = 0$ (d) $s = 10$

B Areas

To find the area of a rectangle
you multiply its length by its width.

area in cm² = length in cm × width in cm

We can use a shorthand for this.
Instead of *length × width* we can write

$l \times w$ or (even shorter) lw.

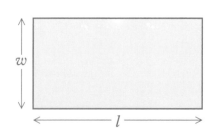

In algebra, when you see two letters together
it means you have to multiply.

So we write $a = lw$
a stands for the area of the rectangle in cm².
l stands for the length of the rectangle in cm.
w stands for the width of the rectangle in cm.

B1 In the formula $a = lw$ work out a when $l = 5$ and $w = 8$.
(Copy and complete the working shown.)

> When $l = 5$ and $w = 8$
> $lw = 5 \times 8$
> So $a = \ldots$

B2 In the formula $a = lw$, work out a when
(a) $l = 4$ and $w = 7$ (b) $l = 3 \cdot 2$ and $w = 4 \cdot 5$
(c) $l = 6$ and $w = 2 \cdot 5$ (d) $l = 3$ and $w = 10$

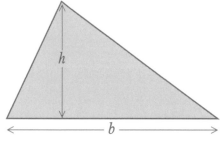

To find the area of a triangle, you
multiply the base by the height and divide by 2.

You can use a shorthand for this.

b stands for the length of the base in cm
h stands for the height in cm
a stands for the area of the triangle in cm².

Then $a = \dfrac{bh}{2}$

$\dfrac{bh}{2}$ means multiply b by h and then
divide your answer by 2.

Here is an example.

The working is done for you.
Look carefully at how we write it out.

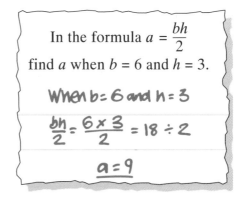

In the formula $a = \dfrac{bh}{2}$

find a when $b = 6$ and $h = 3$.

When b = 6 and h = 3

$\dfrac{bh}{2} = \dfrac{6 \times 3}{2} = 18 \div 2$

a = 9

Now try these.
Write out your working carefully.

B3 In the formula $a = \dfrac{bh}{2}$, find a when

(a) $b = 6$ and $h = 4$　　　　(b) $b = 8$ and $h = 5$
(c) $b = 7$ and $h = 5$　　　　(d) $b = 4{\cdot}8$ and $h = 4{\cdot}5$

B4 Nutty Toys give their workers a bonus.
They use a formula to work out
what each person gets. The formula is $b = \dfrac{yw}{100}$

y stands for the number of years the person has worked for
Nutty Toys
w stands for the person's normal weekly wage in pounds
b stands for the bonus they get in pounds.
(a) Work out b when $y = 40$ and $w = 200$
(b) Find b when $y = 20$ and $w = 150$
(c) Fred has worked for Nutty Toys for 25 years.
His weekly wage is £120.
How much bonus does he get?
(d) Jean has worked for Nutty Toys for 1 year.
Her weekly wage is £80.
How much bonus does she get?

B5 If $z = \dfrac{xy}{5}$ work out z when

(a) $x = 4$ and $y = 10$　　　　(b) $x = 8$ and $y = 2$

B6 If $s = 5ab$, work out s when $a = 4$ and $b = 6$.
($5ab$ just means $5 \times a \times b$.)

B7 If $p = 4sm$, work out p when $s = 3{\cdot}5$ and $m = 12$.

This formula tells you the volume of a cuboid.

$$v = wlh$$

w stands for the width of the cuboid in cm
l stands for the length in cm
h stands for the height in cm
v is the volume in cm³.

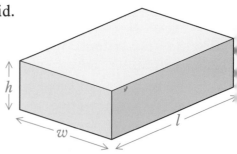

wlh just means $w \times l \times h$.
For example, if $w = 6, l = 4$ and $h = 2$ then $wlh = 6 \times 4 \times 2 = 48$.

B8 Use the formula $v = wlh$ to work out v when
   (a) $w = 3, l = 10$ and $h = 4$ (b) $w = 3{\cdot}2, l = 4{\cdot}5, h = 6$
   (c) $w = 8, l = 12$ and $h = 5$ (d) $w = 0{\cdot}75, l = 2, h = 1{\cdot}5$

This shape is a prism.
It has the same triangular shape
all the way through.
We call it a **triangular prism**.
There is a formula for the volume
of a triangular prism.

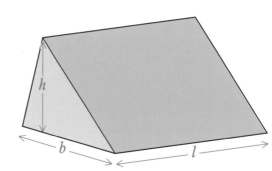

It is $v = \dfrac{hbl}{2}$

h, b and l stand for the height, base and length in cm.
v stands for the volume in cm³.

To work out $\dfrac{hbl}{2}$ you multiply h by b and by l and divide your answer by 2.

B9 Use the formula $v = \dfrac{hbl}{2}$ to work out v when

   (a) $h = 6, b = 8$ and $l = 15$ (b) $h = 2{\cdot}5, b = 4$ and $l = 12$

B10 Use the formula $v = \dfrac{hbl}{2}$ to work out the volume of
each of these triangular prisms.

(a)

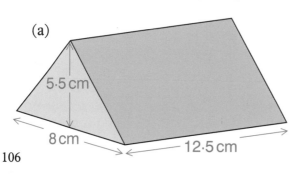

(b)

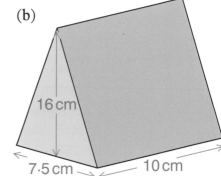

c A mixture

C1 You can use a formula to work out average speeds.

$$s = \frac{d}{t}$$

s is your average speed in m.p.h.
d is the distance you travel in miles.
t is the time you take in hours.

(a) Shelagh runs 18 miles in 3 hours. What is her average speed?

(b) Jan drives 200 miles in 4 hours. What is her average speed?

(c) Meg flies 2100 miles in $3\frac{1}{2}$ hours. What is her average speed?

C2 In the formula $a = ef$, work out a when

(a) $e = 3, f = 2$ (b) $e = 8, f = 6$ (c) $e = 4, f = 5$ (d) $e = 4, f = {}^-5$

(**Hint:** when $e = 4$ and $f = {}^-5$, you need to work out $4 \times {}^-5$.

Look back to the Negative numbers 2 chapter if you need to.)

C3 In the formula $y = xz$, work out y when

(a) $x = {}^-3, z = {}^-2$ (b) $x = 7, z = {}^-2$ (c) $x = {}^-5, z = 10$

C4 If $r = 4t - 6$, work out r when t is

(a) 5   (b) ${}^-5$ (c) ${}^-3$

(d) ${}^-10$ (e) 8   (f) 1

C5 Richton council gives pensioners a cold weather bonus.
They use a formula to work it out.
The formula is

$$b = \frac{at}{{}^-10}$$

b is the bonus each week in pounds
a is the age of the pensioner
t is the lowest temperature that week.

(a) Work out the bonus when $a = 65$ and $t = {}^-2$.

(b) What is the bonus if a is 70 and $t = {}^-5$.

(c) Darby is 80. The temperature falls to ${}^-12°C$ one week!
How much bonus should he get?

(d) Joan is 75. The temperature is ${}^-6°C$ one week.
How much bonus should she get?

Review: tree planting

Stella is planting some young trees.
Each tree is planted in a plastic
tube with a stake supporting it.

1 This scale plan shows the field
where Stella wants the trees.

Measure the dimensions of
the field on the plan.
What are the dimensions of
the real field?

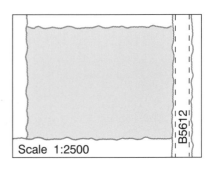

Scale 1:2500

2 Work out roughly what the
area of the real field is.

3 The trees have to be planted 3 m apart.
About how many trees are needed?

2 year old trees	
Oak	34p each
Ash	28p each
Holly	96p each
Walnut	125p each
Hazel	19p each

4 Stella wants to plant 75% oak trees
and the rest walnut trees.
Work out how much the trees will cost.

5 You can buy 50 plastic tubes for £9·50
and 100 stakes for £31·20.

How much will tubes and stakes cost for
the trees?

6 After planting, you spray around the
tree with weed killer.

  (a) What area do you spray for each tree?

  (b) What total area needs to be sprayed
     for all the trees?

*Clear all weeds
from around
tree with weed
killers.*

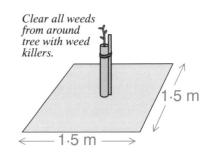

1·5 m

1·5 m

7 Stella estimates that it will take about
5 minutes to plant each tree.
How long will it take to plant them all?

8 Stella can get a grant for planting the trees.
The grant is £200 for each 1000 m² planted.
How much grant can she get?

17 Number patterns

You need Multilink cubes

A Spikes

Each of these shapes is built in the same way.
We call them a **sequence** of shapes.

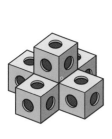

Shape
number 1

Shape
number 2

Shape
number 3

A1 Make the sequence of shapes from Multilink.

Shape number	Number of cubes
1	6
2	
3	
4	
5	
6	

A2 Copy this table.

Count the number of cubes
in each shape.

Write the number of cubes
for shape numbers 1, 2 and 3
in your table.

A3 Make the 4th shape in the sequence.
Count the number of cubes there are in the shape.
Write the number in your table.

A4 You can add cubes to the 4th shape
to make the 5th shape.

(a) Add the cubes.
How many cubes do you add?

(b) How many cubes are there
altogether in the 5th shape?
Write the number in your table.

A5 (a) How many cubes do you need to add to the 5th shape
to make the 6th?

(b) Without making the 6th shape, in your table fill in the
number of cubes there would be in the 6th shape.

A6 You want to make the 6th shape into the 7th shape.
How many cubes will you need to add to the 6th shape?

A7 If you have any shape in the sequence, how many cubes
do you need to add to make the next shape?

A8 Work out how many cubes there will be in the 10th shape.

A9 Work out how many cubes there would be in the 20th shape.

Challenge

How many cubes would there be
in the 100th shape?
Explain to a friend how you can be
sure of your answer.

B Triangles

Shape
number 1

Shape
number 2

Shape
number 3

Shape
number 4

B1 Copy this table.
Fill in the number of cubes
for the first 3 shapes.

Shape number	Number of cubes
1	1
2	3
3	
4	
5	

B2 Make the 4th shape
using multilink.
In the table, fill in
the number of cubes
for the 4th shape.

B3 Add cubes to the 4th shape
to make the 5th shape.
Fill in the table for
the 5th shape.

B4 How many cubes did you add to the 4th shape to make the 5th?

B5 How many cubes would you add to the 5th shape to make the 6th?

B6 Do you always add the same number of cubes
to get from one shape to the next?

B7 (a) Copy and complete this number pattern.

$$1 = 1$$
$$1 + 2 = 3$$
$$1 + 2 + 3 = ...$$
$$1 + 2 + 3 + 4 = ...$$
$$.......................... = ...$$
$$.......................... = ...$$

(b) What do you notice about the number pattern
and the sequence of shapes you just made?

1, 3, 6, 10, 15, 21,...

This sequence of numbers is called the **triangle numbers**.

B8 On squared paper, draw the first seven triangle numbers like this.

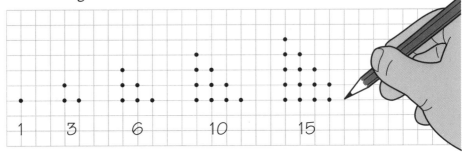

B9 On squared paper, draw the 8th triangle number.

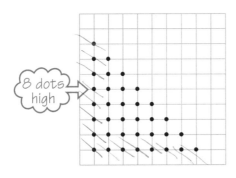

8 dots high

Add dots to turn it into the 9th triangle number.

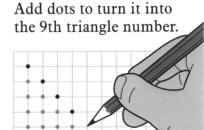

(a) How many dots do you need to add?

(b) How many dots would you need to add to the 100th triangle number to make it the 101st?

B10 This shows the red balls at the start of a game of snooker.

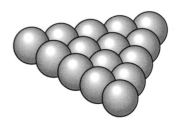

(a) How many reds are there?

(b) Suppose you wanted to use more reds. But you still want them to start in a triangle. How many reds could you use?

c Square numbers

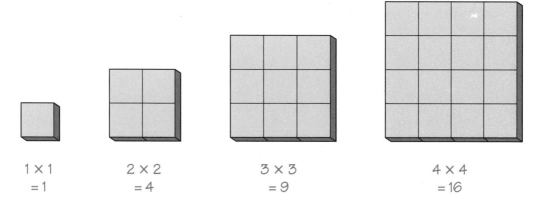

1×1
$= 1$

2×2
$= 4$

3×3
$= 9$

4×4
$= 16$

We call the numbers 1, 4, 9, 16 ... the **square numbers**

C1 Work out the fifth square number.

C2 What is the sixth square number?

C3 Copy and complete this table
of the first 10 square numbers.

1st	2nd	3rd	4th	5th	6th	7th	8th	9th	10th
1	4	9	16						

C4 (a) Copy and complete this number pattern.

$$1 = 1$$
$$1 + 3 = 4$$
$$1 + 3 + 5 = \ldots$$
$$1 + 3 + 5 + 7 = \ldots$$
$$\ldots\ldots\ldots\ldots\ldots\ldots\ldots = \ldots$$
$$\ldots\ldots\ldots\ldots\ldots\ldots\ldots = \ldots$$

(b) What do you notice?

C5 Use multilink to make the 2nd and 3rd **triangle** numbers.
(a) Can you put them together to make a square?
(b) Which square number do they make?

C6 Suppose you make the 5th and 6th triangle numbers.
Which square number do they make if you put them together?

113

D Cubes

D1 Here is a sequence of cubes.
Build each of them with
pieces of multilink.
(The first one is easy!)

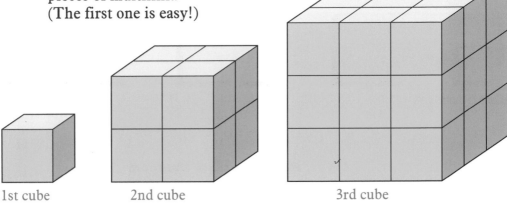

1st cube   2nd cube   3rd cube

D2 (a) How many pieces of multilink are there
in the 2nd cube?

(b) How many pieces of multilink are there
in the 3rd cube?

D3 Copy this table.
Fill in the number
of pieces in the
2nd and 3rd cubes.

Cube number	Number of pieces
1	1
2	
3	
4	

D4 How many pieces of multilink would you need
to build the 4th cube in the sequence?
Fill your result in the table.

The numbers in your table are called the **cube numbers**

D5 (a) Copy and complete this number pattern for the cube numbers.

$1 \times 1 \times 1 = 1$

$2 \times 2 \times 2 = \ldots$

$3 \times 3 \times 3 = \ldots$

$4 \times 4 \times 4 = \ldots$

(b) Carry on the pattern up to $10 \times 10 \times 10$.

E **Shorthand**

Three squared.

The 3rd square number is 3×3.
There is a shorthand for this. We write 3^2.
When you see 3^2, you read it as '**three squared**'.
The 5th cube number is $5 \times 5 \times 5$.
We write this as 5^3 and say '**five cubed**'.

3^2

E1 Write down how you say
(a) 8^2 (b) 9^3 (c) 3^3

E2 We write 'four squared' in numbers as 4^2.
How do we write 'seven cubed' in numbers?

E3 Write each of these in numbers.
(a) three cubed   (b) nine squared   (c) two cubed

To work out 5^2, you work out $5 \times 5 = 25$. So $5^2 = 25$.
Notice that 5^2 is **not** the same as 5×2.

E4 Work out
(a) 4^2 (b) 6 squared   (c) 3^2
(d) 7^2 (e) ten squared   (f) 9^2

E5 To work out 4^3 you work out $4 \times 4 \times 4 = 64$. Work out
(a) 2^3 (b) 5 cubed   (c) 10^3
(d) 6^3 (e) 7^3 (f) 1^3

E6 To work out $6^2 - 5^2$ with most calculators
you need to do it in steps.

$6^2 - 5^2 = 155$
on my calculator.
That can't
be right!

155

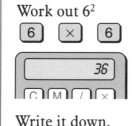

Work out 6^2

[6] [×] [6]

36

C M / ×

Write it down.

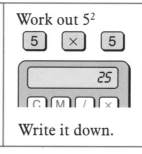

Work out 5^2

[5] [×] [5]

25

C M / ×

Write it down.

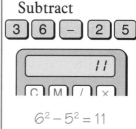

Subtract

[3][6] [–] [2][5]

11

C M / ×

$6^2 - 5^2 = 11$

Work out
(a) $7^2 - 6^2$ (b) $8^2 - 7^2$ (c) $9^2 - 8^2$
(d) What do you notice?

E7 Use the result of question E6 to work out $20^2 - 19^2$ in your head.

E8 Which do you think is bigger, 15^3 or 30^2?
Use a calculator to check.

E9 Suppose you have 400 multilink pieces.
What is the biggest cube that you can make?

E10 Work out $\quad 1^3 + 2^3$
$1^3 + 2^3 + 3^3$
$1^3 + 2^3 + 3^3 + 4^3$
$1^3 + 2^3 + 3^3 + 4^3 + 5^3$
What do you notice?

E11 Write down a number between 100 and 200.
Can you make up the number by adding square numbers together?

You can use the same square number more than once,
but you must not use more than 4 square numbers altogether.

Squared and Cubed

A game for 2 or more players   You need 3 ten sided dice

Take it in turns to roll the dice.
Each player tries to make as many square or cube numbers
as they can from the numbers on the dice.
You score one point each time you use a number on the dice.

For example, suppose
you roll a 1, a 2 and a 5.

You can make a big score from this!

Use the 2 and the 5	25 is a square number (5 × 5).	Score 2.
Use the 1	1 is a square number (1 × 1).	Score 1.
Use the 1 again.	1 is a cube number (1 × 1 × 1).	Score 1.
Use the 1, the 2 and the 5.	125 is a cube number (5 × 5 × 5).	Score 3.
Use the 5, the 1 and the 2.	512 is a cube number (8 × 8 × 8).	Score 3.

So you can score a total of 10 with a 1, a 2 and a 5.

Play the game. The first player to score 20 wins.

Review: chapters 1 and 2

1 This rule tells you the cost of sand, including delivery.

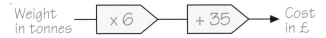

   (a) How much does 10 tonnes of sand cost?

   (b) Write the rule as a formula.
      Use w for the weight of sand and c for the cost.

2 Write each of these rules as formulas.

   (a) $w \times 8 + 20 = C$ (b) $A \div 10 + 15 = P$

   (c) $S \times 5 - 30 = t$ (d) $X \div 2 - 4 = Y$

3 The length of each side of this shape is b cm.
   The perimeter is p cm.

   (a) How many sides does the shape have?

   (b) Write down the formula connecting p and b.

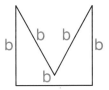

4 This grid shows the
   plans for a new zoo.

   (a) What is at $(1, 1)$?

   (b) What animals are
      planned at $(^-2, 2)$?

5 At what coordinates
   will you find the Bears?

6 The grid lines are
   50 m apart.
   What do you find
   100 m north of $(^-1, 1)$?

7 What are the coordinates of
   the Spider house?

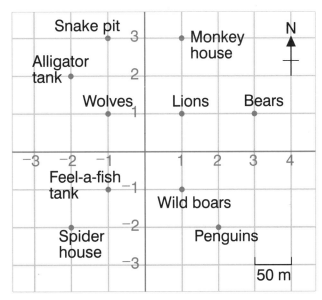

8 How far would a spider have to walk from the spider house
   to the Alligator tank?

9 What can you do at $(^-1, ^-1)$?

Review: chapters 3 and 4

You need compasses, an angle measurer, a ruler and a pencil.

1 This is a sketch of the net of a solid.
On plain paper, draw the net accurately.
Start by drawing the centre triangle.

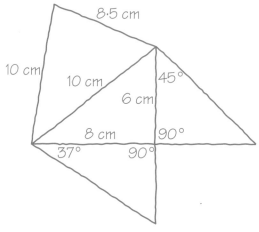

2 Add flaps to your net.
Cut out the solid and make it.

3 Write this recipe out again.
Write it using metric units.

4 The recipe serves 8.
(a) How many lb of lamb
would you need for 4?
(b) How many kg is that?

Lamb carbonnade

6 lb neck of lamb
4 pt beer or stock
2 lb parsnips
1 lb onions
6 oz flour

5 Write this letter out again, using metric units.

Dear Mike – I really think Dad has flipped at last!
He put up a shelf in the kitchen. Jo wanted a small
one, about 12 inches by 6, to put the 3 pint saucepan
on. But no – Dad had to make it 24 inches by 8 inches.
You could put a 100 lb saucepan on it! He couldn't get
any 1½ inch screws, so he used 4 inch nails. I ask you.
When Jo put the 6 pint saucepan on it, all the
nails pulled out of the ...

Review: chapters 5 and 6

1 Work out each of these.

(a) $^-3 + 4$ (b) $^-4 + ^-6$ (c) $^-10 - 2$

(d) $4 + ^-6$ (e) $^-5 - 6$ (f) $^-5 - ^-6$

(g) $^-10 + ^-12$ (h) $^-12 - 4$ (i) $^-8 + 2$

(j) $^-15 - ^-5$ (k) $15 - ^-4$ (l) $0 - ^-4$

(m) $^-16 + ^-3$ (n) $^-18 - ^-8$ (o) $^-10 + ^-10$

(p) $^-8 - 5$ (q) $^-18 + 10$ (r) $17 + ^-8$

(s) $0 + ^-6$ (t) $6 - ^-3$ (u) $5 - ^-5$

2 How many planes of symmetry does each of these shapes have?

A 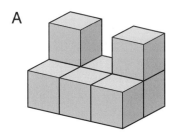   B    C

3 Which of the solids in question 2 have rotational symmetry?

4 Each of these solids has an axis of rotation shown.
What is the order of rotation symmetry of each solid
about its axis?

A 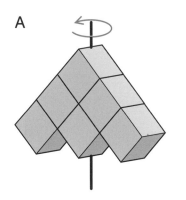   B 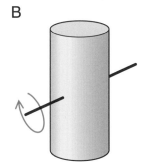   C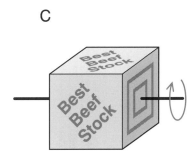

Review: chapters 7 and 8

1 Work out each of these.

(a) ⁻2 × ⁻3   (b) 2 × ⁻5   (c) ⁻3 × ⁻7

(d) ⁻8 ÷ 2   (e) ⁻16 × 4   (f) ⁻16 ÷ ⁻4

(g) 20 ÷ ⁻4   (h) 16 × ⁻2   (i) ⁻18 ÷ 3

(j) 24 × ⁻2   (k) 24 ÷ ⁻2   (l) ⁻24 × 2

(m) ⁻24 ÷ 2   (n) 30 ÷ ⁻5   (o) ⁻36 ÷ 6

(p) 10 × ⁻3   (q) ⁻2 ÷ ⁻1   (r) ⁻1 × ⁻6

(s) ⁻64 ÷ 8   (t) 20 × ⁻2   (u) ⁻20 ÷ ⁻4

2 Here are some angle facts.

Work out angle *a* and angle *b* in these diagrams.

Say which angle fact you used to work each angle out.

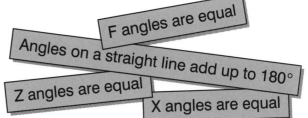

3 Work out each angle marked with a letter.
Copy the diagrams if you want.

Not to scale

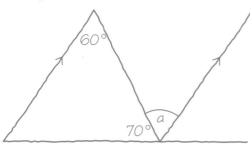

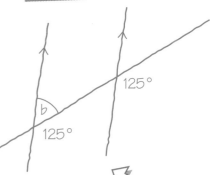

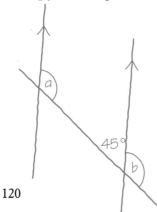

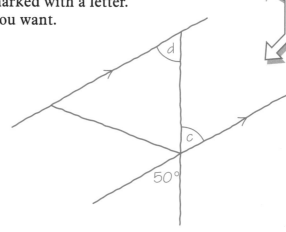

120

Review: chapters 9 and 10

1 Work out the areas of each of these triangles.

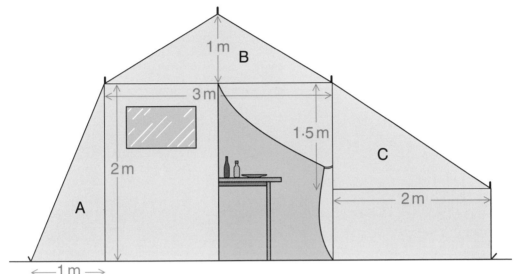

2 This triangle is drawn
to scale.
Measure what you need
and work out its area.

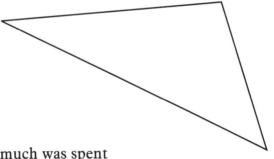

3 The table below shows how much was spent
on different types of advertising in 1991 in the UK.
For example, 27% of the money spent on advertising
went on TV adverts.

Type of advert	TV	Posters	Radio	Papers/Magazines	Cinema	Direct mail
Percentage	27%	2%	2%	55%	1%	13%

(a) Draw a pie chart to show this information.

(b) In Britain in 1991, a total of £2350 million
was spent on advertising!
How much of this was spent on TV adverts?

Review: chapters 12 and 13

1 Here are some angle facts.

Work out angles *a* and *b*.
Say which angle fact
you used to work each
one out.

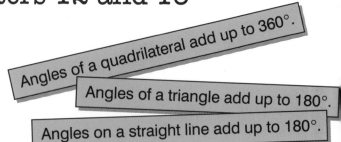

Angles of a quadrilateral add up to 360°.

Angles of a triangle add up to 180°.

Angles on a straight line add up to 180°.

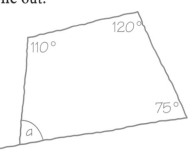

120°
110°
75°
a

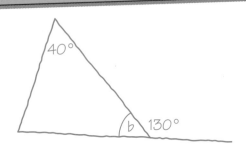

40°
b 130°

2 Work out each
angle marked
in this diagram.
(The diagram is
not to scale.
You cannot measure.)

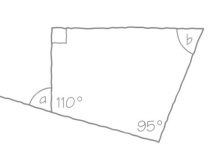

b
a 110°
95°

3 A cube has a volume of 500 cm³.

Copy and continue the working below.
Work out the side of the cube
to 1 decimal place.

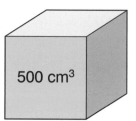

500 cm³

Side	Working	Side too small	Side too big
20	20 × 20 × 20 = 8000		
10	10 × 10 × 10 = 1000		20
5	5 × 5 × 5 = 125		10
		5	

Review: chapters 14 and 15

1 This monument is made of granite.
   The base is a cuboid with a
   prism on top of it.

   (a) What is the volume
       of the base?

   (b) Work out the volume of
       the prism.

   (c) 1 m³ of granite weighs
       about $2\frac{1}{2}$ tonnes.
       How much does the monument
       weigh altogether?

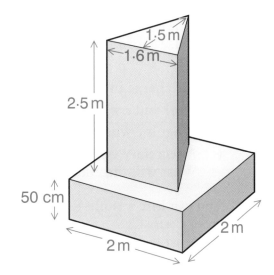

2 Look at the table below.
   There is a tick in the box against 9 and odd
   because 9 is an odd number.
   There is a cross against 4 and odd because 4 is not odd.
   Copy the table. Complete it with ticks and crosses.

	is odd	is a factor of 60	is a prime number	is a multiple of 3	is a square number
4	✘				
9	✔				
10					
12					
13					

3 Copy and complete these factor trees.

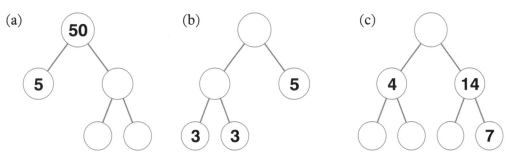

Review: chapters 16 and 17

1 *Man Friday* hire out secretaries.
They use a formula to work out the charge.
$$c = 6h + 10$$
c is the charge in £, h is the hours the secretary works.
(a) Work out c when $h = 8$.
(b) What is c when $h = 24$?
(c) A secretary works for 5 hours.
What will the charge be?

2 You can hire a Rolls-Royce
from *Costly Cars*.
The cost is given by

$$c = \frac{m}{2} + 15$$

m is the miles you drive, c is the charge in £.

(a) What is the charge when $m = 100$?
(b) Work out c when $m = 500$.

3 Look at this sequence of shapes made from matches.

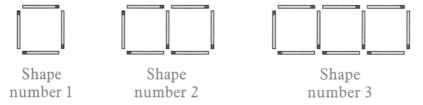

Shape number 1 Shape number 2 Shape number 3

Copy and complete this table.
(Draw your own sketch
of shape number 4.)

Shape number	Number of matches
1	4
2	7
3	
4	

4 What is the special name for these numbers: 1, 3, 6, 10, ...

5 Work out
(a) 6^2 (b) 4^2 (c) $6^2 - 4^2$ (d) $13^2 - 12^2$